Hegel was right

DAEDALUS

Europäisches Denken in deutscher Philosophie

Herausgegeben von
Volker Caysa, Faustino Oncina Coves,
Hans-Martin Gerlach †,
Leon Miodonski, Friedrich Tomberg
und Klaus Vieweg

Band 19

PETER LANG
Frankfurt am Main · Berlin · Bern · Bruxelles · New York · Oxford · Wien

José Porfirio Miranda de la Parra

Hegel was right

The Myth of the Empirical Sciences

Translation by
Eduardo Charpenel Elorduy

PETER LANG
Internationaler Verlag der Wissenschaften

Bibliographic Information published by the Deutsche Nationalbibliothek
The Deutsche Nationalbibliothek lists this publication in the Deutsche Nationalbibliografie; detailed bibliographic data is available in the internet at http://dnb.d-nb.de.

Original title: Hegel tenía Rażón
First Spanish Edition: 1989 (UAM) México

The translation of this book was realized
by Eduardo Charpenel.

ISSN 0938-9547
ISBN 978-3-631-61889-9

www.peterlang.de

Without Male's support, this book
would have not been possible
It is the work of the both of us

Contents

Foreword to the first English edition (2011)

The English translation of Porfirio Miranda's *Hegel was right* is of critical importance for contemporary philosophy as well as for world affairs, since its contribution for the scientific analysis and ethical transformation of present societies is of universal reach.

Miranda's work goes beyond description and commentary of Hegel's thought, by providing a thorough integration into Hegel's propositions of the developments taking place within different disciplines during the last two centuries, and by the systematic reconstruction and amplification of his thesis in the context of XX century philosophy. The result is an illuminating confirmation of the central thesis proposed by Hegel, based on historical events during this period, as well as on Miranda's own arguments.

Following a strict logical and ontological demonstration, which is a necessary condition for all philosophy, Miranda's work provides contemporary arguments for the understanding of the human essence and the State. The importance of revealing the reality of the State is crucial for an objective description, analysis and judgement of the swift transformations of human societies in the XXI century and their corresponding public and civil society institutions.

Human beings are the reality and finality of the State, and it has to be distinguished from government. This Hegelian idea is fully demonstrated in its rational and historical content by Miranda, providing a well supported alternative to the limitations of widely accepted theories of society and culture.

Following Kant and Hegel, Miranda states that human beings are an end in themselves, and should not be considered as means for anything, such as power or the accumulation of wealth. The rational demonstration of this ethical imperative as an absolute truth, has enormous consequences for the present and future human history, challenging notions of the individual, the subject,the nature of ethics and freedom, accepted by the postmodern, self-centered and materialistic theories of society.

The recognition that ethical rationality develops gradually over the centuries lays the foundation for a critical, yet optimistic view of human existence in history. After the demonstration that self interest defines the natural human tendency, Miranda argues over the importance of raising the level of ethical consciousness in all subjects, in order to expand the ethical exercise of individual freedom.

This message is of decisive importance at the present time of world history, when the dominant technological system which is based on the digital convergence, the global scale of production processes and markets, and the unparalleled concentration of wealth and production by large corporations and a handful of individuals, set in motion significant contradictions within the nation states as the locus of democracy and human rights compliance.

According to Miranda, since we are neither good nor human by nature, the process of becoming a human being requires the ethical development of each individual, and the ethical transformation of the social order, as well as of all human beings. The guidance of Miranda for this enormous, yet compelling task, is now open for readers of the English language.

Dr. Alberto Montoya Martín del Campo

Foreword to the third Spanish edition

More than twelve years ago, the Metropolitan Autonomous University (Iztapalapa campus) published for the very first time this work, which goes far beyond a scholarly study of the Hegelian philosophy. Now, after a long time, this work comes again to light, as brilliant and important as always.

Sooner or later, everything that is fundamental becomes a classic, and that is the case of this book, for we can find in it a series of analyses, argumentations and theses which have a significant connection with an entire philosophical tradition of twenty-five centuries that goes back to Plato and that is continued by Aristotle, Descartes and Kant, and is later recovered by Hegel. This tradition stands out for its belief in truth; a belief that is not based on blind recourses or arguments of authority, but rather on demonstration itself. In this text, Dr. Miranda reminds us that there are many things which have been demonstrated for all times, and precisely in that their actuality consists.

It is relevant in many ways that theses which are taken as truths are demonstrated: first, because it makes explicit the departure of irrationality, of the primitive magical thought and of the aesthetic mentality: all subjects which have been addressed in this text, taking off from the fact that we come from the animals and that rationality is not natural to us. Second, it becomes manifest that civilization has to do with the progress of rationality in the world, and that would be impossible without truth and the apodictic character of it. Another of its features is the development in philosophy of what has been expressed in Christian theology "Truth shall make you free" (John 8, 31). But, in what sense can a man be free other that in self-knowledge? (*Nosce te ipsum*) How can we choose without knowing? How can we know without truth? How can we have truth without having proofs?

The denial of truth, as it has been expressed by superficial skepticism, has had on axiological and psychological terms a disastrous effect —one speaks in our time of the Generation X—, for the only thing it has achieved is that lies, mental confusions and contradictions are now surreptitiously considered as absolute truths.

The important thing is that, without truth, no true compromise can exist, and without the latter a fundamental ethical attitude cannot exist either —a fundamental ethical attitude that compels us always to make this world more just and worthy.

Basing himself on his clinical data, post-kleinian psychoanalyst Wilfred R. Bion wrote that truth is the nourishment of the soul, for only by means of it one can modify and improve the mental, emotional and material quality of life among humans. On the contrary, says Bion, lies are venom. Modern skepticism has taken charge of spreading this widely. *Hegel was right* is a powerful antidote against that venom.

In our times of economic globalization, it seems imperative that those of us who are impacted by it are capable of a critical revision of such process, and that is only possible if thought itself is not global or economic. In other words: we are only capable of advancing solutions and of proceeding accordingly if reason itself can be autonomous from the material and psychical pressures which can be exerted upon human beings.

The work the reader has in his hands offers many important subjects: it is impossible to address each of them in this introduction, so I will only mention a few:

1. The difference between science and literature, which consists, namely, in that the former is compelled to demonstrate the truth of its statements, while the latter is not.
2. The lack of synonymy between beauty and truth. It becomes evident hereby that a work or a thesis can be beautiful and pleasant, but there is a great difference between that and something being true.

3. The synonymy between science, philosophy and the common substance, which consists in defining and demonstrating.
4. The non-empiricity of scientific concepts and, consequently, the participation of the subject in their coinage.
5. The impossibility of obtaining *a posteriori* the concept of science, which means it is impossible to obtain that by the means of observation and generalization that scientists commonly employ.
6. The refusal of the widespread Rousseaunian story of natural goodness. The argument we find here is that man makes himself by means of cultural and historical development and not by a purportedly natural determination.
7. Self-determination or freedom which is nothing more than the substance that we call spirit (*Geist*).
8. The ethicity as a self-reflective and constitutive phenomenon of being.
9. The family, the society and the State.

The sequence of chapters in this work offers a transition from philosophy of science and ontology to natural sciences and political philosophy. The link between the different chapters is perfectly logical. Thus, *Hegel was right* is a long and yet rewarding travel that touches not only the writings of Hegel but the most fundamental themes in the history of Western philosophy. One cannot think the way one used to do after reading this book.

The history of Professor Miranda begins in 1924, in the city of Monterrey, Nuevo León, and it is a history full of intellectual and practical braveness. It goes from the powerful critiques of his *Hunger and Thirst of Justice* to his *Marx and the Bible* —which made of him a pioneer of the Liberation Theology— to his demolishing refutation of the positivism of Comte and the Vienna Circle in his *Appeal to Reason*.

A thorough revision of contemporary philosophy would be incomplete without the pages that Dr. José Porfirio Miranda has bequeathed us. Thank you for this valuable heritage.

Héctor Villanueva

CHAPTER I

Science and Literature

It has been a clear ruse of escapism to confuse Philosophy with Literature, as well as to think that the scientificity revindicated by Hegel is a strange genre that bears no relation with the so-called scientific disciplines. The fact that Hegel denounces "sleight of hand and garrulity" in Newton's demonstrations (WL I 277) ought to have induced commentators and scientists to suspect that something without precedents was taking form there.

In the first part of this chapter, I shall focus on aestheticism and other similar gags that have been put on Philosophy's mouth. This discussion will bring us to one of the most important problems of our times and of all times.

Once all the vetoes, which prevent that scientific requirement is even formulated, have been overcome, our second part will address the core of scientificity's exigencies.

1. Literature and caprice

In the latter decades it has been widely discussed, whether Literature must be put at stake. But deep down, it never is, despite certain appearances. Whoever does not *demonstrate* the truth of his claims displays them without a true compromise with what he or she says. He displays them because he 'likes' them. Although the case could be that he doesn't.

Within the discipline of Literature, no author is obliged to demonstrate the truth of his claims, but in Philosophy, everyone is. Otherwise, what we have in our hands is a literary essay:

> Those who think that in Philosophy one can do without demonstrating and deducing proof, thereby that they are still away from the most primary notion of Philosophy; they can speak about anything, but those who want to speak without concept do not have any right of peeping into Philosophy. (Rph, 141A)

In the worst cases, a poet simply lets himself go and speaks out arbitrarily the theses that come to his mind; no one demands that he proves their truth as long as they are beautiful. In extreme instances, the fact that I am a literary writer is a letter of marque, a permit to be irresponsible: since I am a literary writer I can endorse whatever I want; in the end, I do not have to prove it.

In the best cases, the literary writer is convinced of the truth of his theses and would probably like also that others share his conviction. However, he would like to do this not by means of demonstration but by the beauty of them. Even his personal convincement does not depend on any prove but on the beauty and transforming capacity of his own theses and conceptions. Nevertheless, Hegel tells us that "beauty is no yet truth". (WG 601)

It would be a great mistake to think that Philosophy wants to become an inquisitor or that it is inclined towards censorship and intolerance. Philosophy does not appeal to authority but to the consciousness of literary writers themselves, and poses for the whole world a decisive question that, undoubtedly, has been neglected: Is promoting rationality the same as promoting that man *disregards* the truth or falseness of the beliefs and conceptions he adopts?

It is neither a particular false claim nor the circulation of it what irritates Philosophy, but the fact itself that, when adopting a worldview, one considers the question of its truth or falseness indifferently. Such indifference is sheer and pure irrationality. To encourage it is to draw man backwards to an animal stage in which he was controlled by emotions, impressions and feelings, but not by reason. To adopt a theory 'because I like it', to be a materialist or a spiritualist, a liberal or a conservative 'because I like it' is to return to primitive savageness and to be as close as possible to original animality.

By the way, it is not a regression impelled by a recent predominance of an aestheticist criterion: the latter only favors its use, but it does not create the criterion itself. Its cause lays rather in the spontaneous tendency to go back to nature. Rationality is not given to us by nature. What is new, what is extraordinary, what is news in the strong sense of the word, is that the time of rationality has arrived, the time in which it no longer suffices that a worldview pleases, for one demands its truth.

That and no other is the sense of a famous Hegelian thesis that has shocked artists to a great degree: "the spirit has been left behind art" (PG 492), "the characteristic way of artistic production and of its works does not longer satisfy our highest need" (Ästh I 48), "the beautiful days of the Greek art and the golden period of the late Middle Ages belong to the past" (Ästh I 49).

In order to grasp this thesis we must understand the Hegelian news we just mentioned. First of all, we must realize that there *is* news and that "our time is a time of birth and a transition to a new period" (PG 15): today we can resist the siege of a genuine artwork by saying that, after all, it is only Art. One must still find out if its message is genuinely compelling due to its truth: "the impression that artworks provoke nowadays is of a more sober kind; what they arouse in us still requires a higher criterion and a different testimony" (Ästh I 48). The moment of scientific exigency has come; that is the teaching of the *Phenomenology*: "It is this coming to be of knowledge or science as such what is described in the *Phenomenology of Spirit*." (PG 26)

Because they did not understand this, critics thought that the above mentioned Hegelian thesis of Art foretold the very end of it, at least in the sense that there would not be superior realizations than those of the artists of the Greek Antiquity, the Middle-Ages and the Renaissance. The next paragraph would have sufficed to make them see the truth: "Certainly, we can expect further developments and perfections in the field of arts, but the form of Art is no longer our supreme necessity". (Ästh I 170). Our supreme necessity is truth, and therefore science, for science demonstrates how much truth or falseness there is in something. It has taken entire millenniums of maturing and efforts for humanity to reach a stage of rationality which consists in adopting a conception only if its truth is demonstrated; fortunately enough, we have reached that stage.

Philosophy is not against Art; let us say that loud and clear; the truth of the matter is that the exigency of truth is something much more superior and, fortunately again, that it is something we can formulate. There is a possibility that John Hospers may be right:

> Art provides the most intense, concentrated, and sharply focused of the experiences available to man. Because of this, art can have an enormous influence on the tenor of a person's life, more influential no doubt than any particular system of morality. (EB 25, 719, 1)

On the other hand, Hospers himself will not deny that today's relativism can stir the foundations of any masterpiece. Even though Jean Valjean perceives the sublime depths of the duty towards his neighbors and fulfills it by sacrificing everything without hesitation, the skeptic would say: "How foolish, honesty is but a prejudice, that conduct is very beautiful but suits Valejan only, each one has his own tastes, one would have to demonstrate that such conduct is better than its contrary and that it is obligatory." Against relativism the only effective weapon is science, but only science that goes deep down into the depths of truth. On the other hand, due to the intrinsic irrationality attached to this conduct, it is unacceptable that one ignores if a worldview is true or false. It is not enough that an intensively beautiful conduct overwhelms us; it is not absolute unless it is demonstrated. Hegel says "it is useless; we do not longer kneel before (beauty)" (Ästh I 170)

> When there are no reasons, one appeals to feeling. He who proceeds thus must be left speaking alone, since he goes back to the unity of his peculiarity, which must remain unaltered. When he appeals to his own feeling, the interrelation with us is broken. On the contrary, with thinking and concepts, we are in the field of universality, rationality, and we have in front the core of the matter; only that can we understand. (PR I 102).

"The antihuman, the animal, consists in remaining with and communicating only by those means" (PG 56). Aristotle had already said it: "If he makes none, it is absurd to seek for an argument against one who has no arguments of his own about anything, in so far as he has none; for such a person, in so far as he is such, is really no better than a vegetable" (Metaphysics, IV, 1006a 14).

About the Athenians, which are the esthetical people *par excellence*, Hegel made two remarks that provide us with great insight, for they precise the attitude of Philosophy towards Art, an attitude embodied by Plato: "that slanting, that Art as such becomes the highest and hence the content loses interest, belongs to the Athenian people. Plato did not banish Art from his Republic: he only prevented that it continued to be god" (WG 639)

It is unacceptable that aesthetics becomes the supreme criterion, for beauty is not a synonym of truth, and to postulate that something is true because of its beauty would lead us to believe in a preestablished harmony which has been proved false by the experience of many centuries. For that reason Plato says that "the distinction between Philosophy and poetry [...] is an old quarrel" (Rep 607 B) and his remark becomes even brusquer: according to him, "poems are easy to do for he who does not know reality, for it is phantoms (*poiosin*), not realities that poets produce" (Rep 599 A). In contrast to the predominant aesthetic criterion of a society, Philosophy acknowledges that every human being is compelled to seek the truth, since rationality is mandatory to everyone. Probably, to make a genuine poem is not easy, even for him who does not acknowledge the truth, but what matters is the difference clearly expressed by Hegel: "If it was only a matter of stating things, philosophizing would be an easy task" (BS 422). "The philosophical method does not allow mere suppositions; what has value in Philosophy must be proved true, that is to say, its necessity has to be displayed" (Ästh I 65).

For that reason, it seems to me that a methodological focus like the following one (which is typical among Hegelian commentators) jeopardizes every interpretative pursuit:

> The question with which we are faced —and this will be true in all of what follows— is neither whether Hegel is correct in what he says nor whether his interpreters are justified in what they say of him. Rather the question is one of finding out just what Hegel does say and of determining what impact that can have on our own thinking. (Lauer 1982, 2)

To study a literary work with this method is fine; on the contrary, the only thing that matters in Philosophy is to prove whether the author demonstrates or not, i.e. if he is right or not. The impact a work may have on us, "demands still a higher criterion and a different validation" (Ästh I 48). If we do not ponder the rigorousness of the demonstration while examining a Hegelian thesis, we cannot understand its meaning. Hegel warns us that "the demonstration remains within the statement" (GP II 398). Thus, if we do not get the demonstration, we are not getting the meaning of the argument. The aesthetical approach is probably the origin of the accusations of obscurity raised against Hegel. Surely the source of great diversity existing of the 'interpretations' of his Philosophy lies therein.

Whoever does not read the *Science of Logic* as a frontal attack against the deficient scientificity of the disciplines so-called sciences does not make an approach to Hegelian Philosophy as a system that pretends to be true; therefore, one approaches it as an 'opinion', as one of the different worldviews that are exhibited in the history museum for someone to *like* them. Such an approach is aesthetical, not philosophical. How is it possible that these people beared hopes to understand a philosophical work if they did not read it as such? "The spirit that gives life in a Philosophy needs, in order to be manifested, to be borne by a similar spirit." (JS 16) Hegel underlines strongly the following fundamental principle:

"The courage of truth, the faith in the power of the spirit, is the first condition of Philosophy". (EGP 5s).

By saying this we walk deep into a territory planted with mines, for this matter does not only regard aesthetics. The aforesaid museal mentality is nowadays an epidemic among promoters of 'culture', a culture which is vague and without contents: culturalism. Other obstacles are the civilizatory relativism and anti-westernism, for how is it possible that a Western man pretends to tell the world what is true and what is not? That goes in hand with indigenism. And we could not miss also the political dimension under the guise of pluralism and of tolerance with declamatory indignation against those who pretend to be the "guardians of truth". We will examine each of this '*-isms*' and their efforts to shut Philosophy's mouth, until we reach at last the enormous blunder that lies at the bottom of the question.

If we start from the principle that truth is unknowable to the human mind, the possibility of any science is annulled; but a contradiction is simultaneously embraced from the beginning, since this principle is taken as a truth, a known truth. We will get back to this point. But if we start from the fact that truth is knowable to the human mind, *one human mind would have had to be the first one to know it*. We do not posses knowledge by nature, and actually, we depend on other human beings to acquire knowledge. This first human mind would have had to belong to some civilization. To say yes, that truth is knowable by reason, without recognizing its factual knowledge by some individual and concrete mind, is to deceive oneself and believe that one deceives the others. But we are not deceived: it is false that whoever thinks this way holds that truth is knowable: "The incredulity that does not deny the general or the general possibility and yet never believes to have it in any particular case does not actually believe in the reality and truth of the matter." (GP I 500)

To the museal mentality the former philosophies belong to the past; they are things of old times, they do not concern us nowadays, and if they do, it is only out of curiosity or an interest in 'culture'.

> Then we have to deal with things that do not belong to us. One gets into the historical as such, does not spare his work and thinks that something has to be done at last in order to survey the thoughts and opinions of others. One tries thus to do without the thing itself, while being oneself outside. One does without the truth and renounces to understand it. (EGP 281)

> No philosophical system can renounce to the possibility of such endeavor: every one of them is susceptible of being treated historically. As every living figure it belongs at the same time to the (world of) phenomenon, a Philosophy as phenomenon surrenders to the very instance that is capable of transforming it into a static opinion and something that belongs in the first place to the past. (JS 15s)

It is a 'collection of mummies', as Hegel says right away. Underlying such collectionism we find the conviction that truth is not knowable, for a true thesis cannot be neutralized as if it did not concern us. If a thesis is true it becomes a bomb that explodes right in the middle of our contemporary world and shakes the foundations of our prevailing convictions.

Whoever studies Philosophy from a 'cultural' point of view,

> stands firmly in his indifferent position towards truth and preserves its independence, whether it accepts or rejects opinions, whether it remains indifferent or not; the philosophical systems are to him nothing more that

> opinions, and such accidental things do not do him anything; he does not know that truth exists. (JS 16)

The first thing to point out is that culture means to 'cultivate', but whoever promotes indifference regarding truth cultivates irrationality, just as he who foments the adoption or the preservation of a worldview because it is pleasant or for any other reason that is not consistent with its trueness.

But the most decisive thing to emphasize is that, with such an approach, one can exactly perceive what is philosophical about philosophies. To the best result such a beginner's approach could lead —and this is still dubious— is to determine what Hegel says, but not *why* he says it; now, that *why* is what is philosophical. "Those people see everything in a Philosophy, but they leave aside Philosophy itself" (GP II 380). That which is philosophical is demonstration. "In Philosophy, in my *Science of Logic*, what is done is *demonstrating*, not exhibiting" (BS 413). The true situation is this: pluralism may seem very tolerant, neither eurocentrist nor partidarian, but the only one forbidden to speak in its international assembly is Philosophy: Pluralism takes Philosophy as Literature, depriving it from its sting of truth and, ultimately, suppressing it. Pluralism is deliberately one-armed: Philosophy is forbidden to speak. Truth may be pursued as long as no one ever finds it, in other words, as long as it is never demonstrated and as long as no one ever makes Philosophy.

Aside from this flagrant intolerance, it is necessary to emphasize this on a political level. Pluralism does not want the existence of absolute truths because if they are absolute they are 'imposed', and this seems incompatible with tolerance. But, first, we do not suggest that they are imposed by the government; we address rather one's duty of knowing the truth. Second, the preposition 'one plus one equals two' is an absolute truth that imposes itself, and nevertheless that does not seem a violation of tolerance to pluralism. Third, and more importantly, tolerance itself needs to be grounded; otherwise, the defenders of intolerance would have as much reason —or unreasonableness— as we who defend tolerance; and the only possible groundwork is that every single human has infinite dignity and hence cannot be unwillingly subjected to certain convictions or beliefs. However, that every human has infinite dignity is an absolute truth. That being said, how can pluralism reject absolute truths in the name of tolerance?

Lovers of 'culture', in such a vague and indifferent fashion, believe that any cultural contribution can be added accumulatively in the mind of people or individuals. Since the problem of truth is indifferent to them, they disregard the evident fact that a contribution could entail opposition to another contribution and destroy it. They figure that anything can be added to anything. In reality, it may happen that both things get subtracted. Cultural works frequently inculcate convictions and judgments of value that are incompatible. True culture cannot do without the contents of truth, for that would be a heartless act that embraces a flagrant contradiction. And this is what is important: to impose contradictions on a mind is not to cultivate it but to demolish it.

> That spirit is the absolute and general alienation of reality and thought: pure culture. What one experiences in such a world is a lack of truth of the essences of power and wealth, of their determined concepts, of good and evil, of the noble and the base; all these elements barter each other and each of them is contrary to itself. (PG 371)

> This eclectics are, on the one hand, those general and uncultivated men, in whose minds the most contradictory representations find their place, without ever gathering their thoughts and realizing their contradictoriness; on the other hand, those cultivated ones who proceed consciously so with the belief of doing the best, taking from each system what is good, as they call it thus they look for a sample of diverse thoughts, in which they sum all the good except coherence in thinking and thereby thinking itself. (GP 431s).

They have thoughts but they do not think; they have them abstractly, pinned down with needles as if they were collectible items, but they do not understand them. When it comes to Philosophy, they miss what is philosophical, for congruency lies in demonstrations, and that is the core of Philosophy.

If we take a closer look to the question, a promotion of 'culture' that does away with truth or falseness cannot speak either of *promotion*, or of the *progress* of man, or of the perfectibility of the human being. It may speak of motion, movement, or change; but that remains dubious, for it would go around in circles and there would be no change at all. If one does not know the goal towards which movement strives, one cannot speak either of progress or of change. 'Movement for change', as a popular Mexican slogan went, is something simply ludicrous.

"In reality, that perfectibility is more or less the same as the mutability in general; it does not have a beginning or an end; the best, the most perfect towards which it must strive is completely undetermined" (VG149s). "In those conceptions progress has in general the form of the quantitative [...] one can chit chat endlessly without getting to any determination and stating something qualitative. [...] The quantitative [...] is that which lacks thought." (VG 150)

Pluralist and unengaged culturalism, in so far it neglects what is philosophical when it peeps into the history of Philosophy and disregards what Philosophy has *demonstrated*, gives always into the temptation of believing that the history of Philosophy is a chaotic parade of systems, or that each system dismisses and ruins the previous ones. It joyfully infers that it is not worthy to deal with Philosophy. It is unbelievable: it focuses on what is not philosophical and extracts conclusions against Philosophy. But what is demonstrated cannot be contradicted: "Those people see everything in a Philosophy, but they leave aside Philosophy itself" (GP II 380).

> For this is usually the language of those who not only justify to themselves their own hatred against Philosophy, but also they take pride in saying that the philosophical systems contradict each other a lot and change so often that it can only be prudent not to deal with them. [...] What has been has been and is in fact transitory are the various attempts of philosophizing without Philosophy and of wanting to have a Philosophy without Philosophy (NH 438s).

"Since there are many philosophies they conclude that there is none." (EGP 272).

Not yet dealing with contents, Hegel responds to them: "In the first place one must say: there is only *one* Philosophy. This has a formal sense, since each Philosophy is at least Philosophy" (EGP 123)

The aforementioned skeptics disregard the only essential thing: the fact itself of the existence of Philosophy, the historical and consistent wonder in which the human reason, after millenniums of irrationality in the adoption of worldviews, has finally decided to seek, definitively and demonstratively, which of them are truth and which of them are false. "The knowledge of truth", so calls Aristotle, Philosophy (*Met* II 993b20). In this sense, Philosophy is only *one*. Philosophers did not exist first and, in order to lunge with prestige their trade, showed up later saying that Philosophy is the profession of finding the truth. No. In one moment of history, human reason had to protest against the above mentioned irrationality; and that moment and the corresponding activity of reason is called Philosophy. The successive linking of systems that there has been is amazing. Of course, 'coffee philosophers' are excluded from the list of genuine seekers of the truth, as well as those who practice a "hazardous philosophizing", as Hegel calls it, by elaborating dilettante and rabble-rousing writings, "often is only chit chat and whimsical opinions what they call Philosophy" (EGP 123) "many philosophical writings limit themselves only to the prattling of feelings and opinions" (EPW 14A).

> Thousands who deal with trivial things have been forgotten; only one hundred names have been preserved. The Mnemosyne of universal history does not confer the unworthy with prestige; just as happens with the deeds of the heroes of the outward history, the only heroes that the history of Philosophy acknowledges are the deeds of the thinking reasons. Those are our object. (EGP 124)

"Philosophizing without a system cannot be scientific" (EPW 14A). In regard of the contents, to show that Philosophy has been *one* and that it is false that some systems displace the others, constitutes the content of this unmatchable Hegelian work called *The History of Philosophy*. To put it briefly, without raising a veil of mystery over the question, we must say that each philosophical system employs concepts less abstract and hence better defined than those of former systems; consequently, it sheds light on more problems and with more concretion than what other philosophies did:

> The beginning as a beginning is simple i.e. the abstract. What is first is actually the rudimentary, that which has not been developed. The principles of the most ancient Philosophy are therefore completely abstract. The most abstract is the easiest: an acute intelligence finds that quickly. (EGP 279)

"In the history of Philosophy the goal is the development of thought. One takes sides in favor of this." (EGP 283) As thought develops, the concepts it uses become more capable of depicting reality because they are more concrete. If we take the principles of the different philosophies in their original abstraction, they overthrow one another; but if we attend to the reality that they cumbersomely tried to point out, that does not occur:

> For instance, the atomistic Philosophy adopts the principle according to which the atom is the absolute; the atom is indivisible, the one —in its ulterior determination, the individual, and, even more determined, the subjective. I am also one, one individual; but I am subject and, hence, spirit. Nevertheless, the atom is the being-for-itself completely abstract, the pure one; [...] The spirit is, of course, also one, but is not the one within such abstraction. (EGP 128)

This subject is extremely important. Physics have not even noticed that the exact translation of *átomon* to latin is *individuum,* and, of course, they have not asked themselves what is the original meaning of this word, i.e. what is the origin of this concept. Empirical experience does not show us any atomic reality, for every sensible data is extended and has parts, and hence the origin of this concept cannot be empirical but introspective: what this concept signifies is what we perceive by means of introspection: the spirit. "The atom and the vacuum are not things of experience" (GP I 359). Who knows what Empedocles, Leucippus and Democritus *imagined* when they used the term atom? However, what matters is not what they imagined but what they said. To be sure, they did not imagine an atom, for the simple reason that the atom is not imaginable; everything that we imagine is extended, and consequently, divisible. They could not have been referring to what they imagined or to what they saw, because an atom is neither imaginable nor visible. When they said that the absolute was an atom, that is to say, something individual, they had all the reason in the world, and the Philosophy of Plato, Aristotle and Hegel did not displace this thesis; it only made it more explicit, exact and concrete. Materialism has never been able to define its terms, and hence it has not been able to declare its grounds without incurring in blunt spiritualism.

The works of the history of Philosophy are not commonly read with the exigency of strict rigor, with the exigency of scientificity, as Hegel reads them. One last and brief example of this: we will see in the last part of the present chapter what is nowadays manifest and evident to intelligent scientists and to anyone who reflects a little, namely, that science cannot trust immediate observation, i.e. empirical data. Immediate observation tells us v.g. that earth is still; but nowadays we know that earth travels 30 km per second. Our unbendable sensations tell us that there is an 'up' and a 'down'; but nowadays we know that this sensation does not correspond to any reality at all, for people in the Patagonia call up what we call down. And this was the first discovery of Philosophy since its birth with the Ionians, the Eleats and the Pythagoreans. As Hegel says about Thales of Miletus: "Here we are introduced into the distancing of what is in our sensible perception, the distancing of that immediate being —a retreating from that" (GP I 203). When the founders of Philosophy affirmed that the only real thing in the world was an underlying element, they were distrusting the testimony of sensation, as if sensible data were only apparent: "one needs great nerve of spirit in order to put aside that richness of the being immediate of the natural world and to reduce it instead to a simple substance that keeps itself as such" (GP I 203). For the Eleats it was the "being", the one; for the Pythagoreans it was the numbers; for the Ionians it was some other element. But that is incidental and marginal: the central thesis is the difference of the empirical data, and this has been confirmed, not displaced, by the subsequent development of science.

But let us get back on track. As a matter of fact, the apparent displacement of some systems by others in the history of Philosophy was a mere pretext of aestheticist culturalism for not taking Philosophy seriously. "The only consequent means against reason is not to deal with reason at all" (WL II 369).

The predominance within a society of the aestheticist criterion in the choosing of a worldview is a demagogy foretold by Plato, who accuses poets and orators of being 'adulators' of the people by giving them what they 'like' instead of what they need (cfr. *Gorgias* 502 B y E, 463A *et passim*). It is perplexing that a diagnose so certain and profound, masterfully expressed twenty-four centuries ago, whose pertinence is even bigger today than in those times for the reason I will immediately expose, does not hold the attention either of political thinkers, sociologists or intellectuals. The remedy to this is not governmental intervention. Undoubtedly, men in our society, with a certain sense of responsibility, must do something about this, given that the predominance of pleasantness has created an effective and unbendable mechanism of repression —as I mentioned before in regard of rebellious, tolerant pluralism. Since only 'pleasant' writings have commercial success, and negotiating editors (who are the majority) only publish what brings them money, the immediate result is that Philosophy is censured... unless it becomes Literature. If Philosophy accepts the rules of this game, its message would not convince because it demonstrates but because it pleases; it would precisely reinforce the irrationality which it must attack.

Men of letters and their editors blame corny soap-operas and yellow journalism because, in their opinion, both things distort the taste of the public and make it incapable of enjoying good Literature. But they do exactly the same in regard of Philosophy, only that in this case something much more important is at stake: the rationality of man. The repressive artifact shuts its doors: the macabre press and the non-sense of novels can intend to challenge us, if we are smart enough, to hit the mark with the taste of people just exactly as they do.

Furthermore, both Literature and pasquinades can give a political tone to the discussion by saying that hitting the mark with what people want is democracy, while pretending to provide them with what they need —instead of with what they want— is totalitarism. In a very similar fashion, vulgar Marxism holds that only the people know what they want. Plato was right when he said that in all this there was too much demagogy involved. This allows us to get at last to the bottom of the problem we have discussed so far, and with respect to which, what Hegel and recent Anthropology have to say, acquires the importance of a revelation.

Probably the biggest mistake that has carried the most unlucky and disastrous consequences over the past two centuries is to believe with Rousseau that man is good by nature. I say 'believe' because Rousseau did only not provide any demonstration of such a grandiloquent claim —a claim which is irreconcilable with the Western intellectual tradition— but practically did not even attempt to provide it. Charron, Buffon and Kolben came before him but they did not have any market echo; it was the first successful *Discourse* of Rousseau what gave this prejudice the enormous popularity it now holds. It besieges liberals, leftists, and even theologians; nobody dares to say the terribly unpopular truth that man is evil by nature, despite it would do a lot of good to our contemporary world. It is the turn of Philosophy to take some action, not by default, but because originality, popularity, commercial success, the mainstream, etcetera, are things of which Philosophy could not care less.

At first, theologians opposed to Rousseau's mistake. But unpopularity is an annoying companion. In particular, the unpopularity that blocks missionary enterprises: to say that man is evil by nature implies that one only becomes good by means of true civilization, something that sounds as intense eurocentrism and seems to be offensive to indigenous cultures. Theology ended up making original sin more subtle and, consequently, more anecdotic, as if it were something committed by someone else, or something that does not discriminate any addressee of learning and does not offend anyone. To blame it on someone who did not perform it would be an arbitrary deed in which one evidently does not have the slightest trace of responsibility; in addition, the regenerative work would consist in the hilarious process of convincing the imputer to stop imputing things. Amidst such subtleties, the opposition to Rousseau ended up fading away. One of its residues has been civilizatory relativism, since if man is good by nature any culture produced by man is good. Other remains have been indigenism and the awareness of those who only look to please or earn money with their writings, for if man is good by nature, his natural tastes and spontaneous preferences would be so too, and he who flatters people and is condescending with them does not do anything wrong.

This third consequence, although historically and logically inevitable, would have outraged Rousseau, but that should not surprise us since logic, as we said before, was not one of his strong points. Let us quote these valuable lines: "There will always be this difference, he who makes himself useful works for others, and he who only thinks in making himself more pleasant, works only for himself. The adulator, for instance, does not spare efforts with the purpose of pleasing, and yet he only does wrong" (Rousseau 1964, 74). This quotation is also pertinent "Truth does not make one rich" (ibid. 371). And from the III book of Émile: "he who rules the folk is not who resembles it". These are sturdy remarks from a very peculiar man.

In order to refute the central non-sense of Rousseau, indigenism and the other two consequences, one would only need to repeat today with Darwin something that sounds rude but which is essential: we come from animals! It is a clear symptom of passion and limited lucidity of indigenisms the fact that they have forgotten (in their explicit or implicit appraisal of their turn to the origins) that we were animals in the beginning. Civilizatory relativism is untenable if the quality of a culture depends on the degree by which man is able to draw away from animality.

To put it in a word: by nature man is not even good; by nature he is not even a man.

Independently from Darwin and Hegel, recent Anthropology, in a detailed study of all the human groups that inhabit our planet Earth, has arrived to a conclusion which our literary writers have not been able to assimilate. I quote Leslie White, one of the most acclaimed modern anthropologists, who is a convinced atheist:

> It used to be said that beneath the endless variety of cultural expression lay a common constant factor. This was called *human nature*, and the omniscient layman was wont to say, as if making a pronouncement, that 'human nature is the same the world over'. But his universal constant is not *human nature*; 'human nature' is *cultural*, not 'natural'. There may well be, and we think, there probably is, a common biological factor underlying all cultures. But it is an animal, a primate animal, factor. Culture, and it alone, can make it human. (Rossi *et al.*, 1977, 250).

One and a half century before, by means of an analysis much more profound and evident in comparison, Hegel had said: "Man achieves what he has to be only by means of culture, by discipline; what man immediately is, is only the possibility of being, that is to say, the possibility of being rational and free, only the destination and duty to be that" (VG 58).

> The soul is spirit only through suppression of the natural will and the appetites. This happens by subjugation under the ethical, by getting used to the ethical as a second nature of the individual; that is, in a word, the work of education, of culture. (PR II, II 178).

"The natural is rather what the spirit has to suppress" (GP II 107). "We used to begin the fiction of a natural state, but that is not the state of the spirit, of the rational will, of the animals between themselves. The war of everybody against everybody is the true natural state, as Hobbes very well remarked" (GP II 108). "The nature of man consists in that he is not what he ought to be" (PR III 109).

"The exigency directed to man is not to be as a natural will, not to be as he is by nature" (PR III 107).

Its true nature is to abandon his immediacy and to look at it as a mood of being in which he must not be. Man as an immediate and natural man has to look himself as somebody who is like he must not be. This has been expressed thus: "man is evil by nature" (PR III 106); "evil is no other thing that the penetration-in-itself of the natural being of the spirit" (PG 539).

> And by the way the natural will is not the will as it ought to be, for it must be free, and the will of the appetites is not free. By nature the spirit is not as it ought to be; it can only be that through freedom; this is exposed by saying that will is evil by nature. But only man is to blame in so far as he remains in his naturalness. (PR I 275).

"The fundamental principle is that man is not a natural being as such; he is spirit, not an animal." (PR, II, I 27)

The principal mistake of Rousseau as well as of civilizatory relativism, indigenism and the eased consciousness of those who earn money pleasing natural 'tastes' fall within what Voltaire expressed in a sharply and incisive judgment:

"On n'a jamais tant employé d'espirit á vouloir nous render Bêtes".

This fulminating line is contained in a letter of August 30, 1755, addressed to no other than Rousseau 1964, 1369). The word-game with *spirit* and *béte* is not translatable into English.

Let us not stop to discuss the argument of those who against Western culture say that other cultures are closer to nature; they obviously demonstrate the opposite of what they intend. Let us rather consider civilizatory relativism and indigenism on a pure logical and cold level.

It is clear that if one professes an axiological relativism, according to which the word 'valuable' does not have sense, the problem is entirely different. I have refuted this doctrine in my *Appeal to Reason*. In the present context let us only say that axiological relativism means to hold that a society cannot be more human than what it already is —something which would not be supported by any nation on this earth. Civilizatory relativism and indigenism want to be based on the sentiments of people.

If axiological relativism is excluded, whoever says that all cultures are equally valuable does not realize that he is advancing the most implausible thesis one could possibly imagine. It would be a miracle, an authentic wonder, that the diverse human groups, which have developed in cultural processes of the most different kinds and have overcome conditioning factors and obstacles of every different sort after being millenniums by their own, have reached today exactly the same result and degree of humanity. The other logical option would be to accept that the results existing today, obtained by different cultural processes, are differently valuable. But since it is ruled out to understand by different the cultures in themselves, this difference has to be explained by biological factors, that means race, and so it seems that indigenists are racists, and worst of all, racists with their own people, unless they embrace axiological relativism, which is for them out of the question —unless they choose the absurd and characteristic regionalism typical of little villages, according to which their piece of land is the best one in the world.

By the way, we could understand *one* indigenism of this sort —but not without a certain smile—, for *a priori* only one of the existent indigenous cultures could be the most valuable; but that is always said from inside hermetism, any comparison with something beyond their piece of land being forbidden. What is complete non-sense is a recent 'universal indigenism' which implicitly exhorts each indigenous culture to feel the best of them all. This amounts to free oneself completely from logic and the duty of knowing the truth.

The possibility that each of the cultures of our planet has arrived to the same stage is a 'superastronomical improbability', to use the expression of biologist Manfred Eigen; in plain words, it is something simply impossible. But, mathematically speaking, it is also a superastronomical improbability that *two* of them have reached equally valuable results. Looking at things with plain objectivity, only one of these cultures can be the most developed in terms of humanity and rationality. I do not know what indigenist spokesmen are so scandalized about.

The indigenous have been the most exploded ones; practical and affective preference for them is not only justified but obligatory; however, only as long as affection does not turn the lights out and produces an ideology that does more against indigenous people than in their benefit.

Every ideology, whose effect is to forbid that a human group incorporates the most advanced level of authentic humanity and the adequate civilization without which its seed cannot germinate, is an ideology that works against a human group by means of adulation. The worst enemy of people is he who flatters them. He who does that betrays them by saying 'you are fine, keep things going that way'. He deprives them from the shame of being savages, something that would be their salvation. The spoiled yearning for the origins protects semirationality by defending it against every 'intrusion' of full rationality. It is irrational to hope that this exiguous and rustic culture, which vegetates at the margins of true and compelling civilization, will gradually reach the best degree of humanity there is now. To begin with, that would mean to condemn to a paltry level of humanity future generations of humans that will not see that happen. Besides, even implausibly supposing that this is possible, when that time comes, universal civilization would reach new heights of true humanity, from which the culture 'protected by indigenists' will be margined.

By the way I would like to say the following. Superficial leftism has contributed to reinforce the paralyzing Rousseaunian conviction that man is good by nature, because if they blame perverted capitalism for the evils of man, it follows that, if capitalism did not exist, man would be good; when in reality what capitalism has made is to disloyally cultivate what man was given by nature, i.e. animality. Besides, at least in Latin-America, it is obvious that the fomented 'class mentality', 'the class sentiment', has been a permit of rusticity and lack of culture, a license for us to be as fool as always. Class mentality should not be confounded with the struggle of classes; without the latter it is impossible that the world arrives to justice. The mentality of classes favors shamelessness; the struggle of classes favors communism.

2. Science without dogmatism

The time has come, we said with Hegel, in which all these dilettantisms and irrationalities come to an end: the time in which man demands a scientific demonstration in order to adopt a theory or a worldview. From now on science has to be the criterion. "The inner necessity that truth must be science lays in the nature itself of the knowledge, and the satisfactory explanation of it is the exposition of Philosophy itself" (PG 12).

Such an exigency must be taken literally serious; it all depends on it, especially, the understanding of Hegel's message: "one of the weak points of our times is not to be able to deal with greatness, properly speaking, the excess of the exigency of the human spirit, the feeling of being overwhelmed and the coward retreat from the enterprise" (GP II 20).

To start with, science is not belief, it does not accept anything by appealing to authority: "What is true has its root in the spirit itself and belongs to its nature; every authority is thereby denied" (GP II 44). But we would incur in the same mistake if we take by authority the concept of science held by some people who nowadays call themselves scientists. First, how could we know if they are true scientists or not? By the public opinion that declares them as such? That would be accepting something by authority, which is unscientific. Second, even though we knew that those persons were scientists, how could we know that the act by which they define science is scientific? We cannot *assume that*: "First, we cannot make suppositions; this is a big principle which is extremely important" (GP III 128). If we start making suppositions, someone else could make contrary suppositions and all would depend at the end of the day on one's whim —something that is the negation itself of science. If scientists themselves say that the act by which they define science is scientific, we would accept by their authority the scientificity of such act, which is an invalid procedure. Furthermore, we would not know if the second act by which they declare scientific the first one is also scientific, for not all acts of men are scientific. They sometimes speak of baseball, sometimes of politics, sometimes of good wines, etcetera.

We do not need to stop here to say that to accept a university degree as a demonstration of the scientificity of a person would be to accept something by authority. First, we know that there are charlatans with a university degree. Second, we would not determine whether the act by which the university granted him the title was scientific.

There are people who, neglecting that which is unscientific to accept a concept of science by simply saying 'I was told so', are still determined to obtain by those means the concept of science they are going to adopt. It is of utmost importance, both to systematic thinking and to the historical moment that the world is going through, to realize that the concept of science cannot be obtained *a posteriori*. It is impossible to obtain it by a generalization or by an induction of the particular acts of science. Scientists and commentators have been reluctant to pay attention to the passage (WL I 23) in which Hegel affirms that determining 'the concept itself of science' is a task of Philosophy. But it is neither arrogance, nor a wish of Hegel, nor a thing that *should* happen (i.e. something optional). It is simply a fact. It is always a Philosophy or a pseudo Philosophy in the head of the person who calls something scientific what determines that he does so. Without a previous concept of what scientific is, we could not know which of the innumerable human acts in the world constitute the reduced group of which one would have to extract, by means of generalization, the definition of science. This concept has to exist in our heads *before* we start the evaluative quest of scientific acts or actions; otherwise, we could not know which acts or actions we ought to examine. It is thus a philosophical concept what defines this despite the negligence of some men of science who claim the opposite.

Even though laymen and some scientists have unscientifically obtained from others the definition of science, this same definition formulated by scientists is a philosophical concept, which has not been obtained by an *a posteriori* induction. Physics do not define as physics what is scientific; they do this as philosophers. Hypothetically speaking, physics could obtain *a posteriori* the definition of Physical sciences by generalizing their own actions; but when they speak of science they do not only refer to their science; on the contrary, they refer to science as such, so that if the particular discipline called physics seems to them scientific, it is because it matches the concept of science as such, which was not obtained *a posteriori* by the observation of the actions of physics, for not all the actions of physics seem to them scientific. Instead, they select which of these actions are scientific by an *a priori* criterion. A scientist would deceive himself if he believes that he obtains *a posteriori* from his own individual praxis the concept of science, for neither all of his actions are scientific nor he holds them to be so.

For instance, when someone conceives the idea (which, by the way, is wrong and mistaken as we will see) that only the act of empirical observation is scientific, he would only judge as scientific his own or other's actions which consist in empirical observation; but that idea is of a philosophical kind and has not been obtained by empirical observation.

If all the scientists of the world could gather and settle on a definition of science, they would achieve this by means of philosophical reasons, not by observation of the things which scientists have done so far, for not everything that scientists have done is scientific.

Einstein's claim that "it is easier to denature plutonium than it is to denature the evil spirit of man" (Schilpp II 1970, 655), will seem scientific to some scientists and unscientific to others, depending on the philosophical concept each of them has of science. If it were a matter of making an induction, some would think that this statement should be included in a scientific data-base, but others would disagree. This makes clear that the intended generalization comes in rather late, for the definition of what is scientific does not only preexist; it is a condition of possibility for the aimed generalization.

The same would happen with the thesis of Einstein contained in his letter to Max Born of December 4, 1926: "The [quantum] theory yields much, but it hardly brings us close to the secrets of the Ancient One. In any case, I am convinced that He does not play dice." (in Jammer 1966, 358, No. 128). 'Some scientists' opinion would be that speaking of God is unscientific; others would think the opposite. Acclaimed physic Freeman Dyson states the following: "When Thomas Wright, the discoverer of galaxies, announced his discovery in 1750 in his book *An Original Theory or New Hypothesis of the Universe*, he was not afraid to use a theological argument to support an astronomical theory" (1979, 245). That is the same thing Einstein does in the above quotation, and the same that Copernicus, Galileo, Kepler, and Newton frequently did. If some scientists of today think that Einstein is not scientific in saying that, they could be right; but they evidently do not think that in function of a concept of science obtained from what scientists actually do. The *a posteriori* character of the concept of science is a persistent illusion which has a lack of reflection as its cause.

Arthur Rosenblueth still falls recurrently in this illusion in his book called *The Scientific Method*, in which he proposes to obtain the concept of science "examining what the scientific treatises may have in common" (1984, 8). It is unbelievable, for Rosenblueth first excludes, having previous conceptions of what science is and what is not, all the parts of the treatises in which descriptions, systematizations, mediations, explanations, and predictions appear. For instance, he excludes explanations by means of these arguments: "Essentially, to explain something to someone is to provide him with a subjective satisfaction which is only incidental to the purposes of science". Regardless if what he says is true or false, here we have a theory of explanations whose *a priori* nature is obvious. We also find here a theory which deals with the relation between explanation and the concept itself of science —a concept that Rosenblueth apparently wanted to extract *a posteriori* from the common elements of scientific treatises.

No less *a priori* and of philosophical nature is the argument he uses to exclude knowledge as the essential core of science:

> The notion of knowledge is subjective and has various meanings. In order to determine the meaning that corresponds to science we may recourse to other criteria or we include them with a word which, ultimately, will be equivalent to the obviously circular statement according to which scientific research pursuits scientific knowledge.

That the notion of knowledge is subjective is a thesis that Rosenblueth evidently did not read in the treatises he endeavored to examine; the rest of his argument is primarily a discussion on logic, and therefore it is not *a posteriori* either.

His other arguments are less flawed, but they are all *a priori* and philosophical. Believing he has extracted this from his analysis on the treatises, he finally draws this conclusion: "we can say that science looks for abstract or theoretical models, which accurately represent the functional relations that exist in nature". Having excluded everything *a priori* that he did not like, he ends up only with what suits his taste. If one objected him what he reproaches to explanations, namely, that those abstract models only provide a subjective satisfaction (which is incidental and unnecessary to science), the discussion would have to be carried through on a philosophical basis, and there would not be any way to solve the problem, for in the existing scientific treatises we find *a posteriori* descriptions, measurements, systematizations, and predictions, as well as the knowledge and the conviction of knowing reality. For instance, Einstein says: ""...behind the tireless efforts of the investigator there lurks a stronger, more mysterious drive: it is existence and reality that one wishes to comprehend." (Schilpp II 1970, 400).

Since only philosophical arguments can determine whether the intention to gain knowledge is essential or inessential to science, the thesis of Hegel proves to be right in the sense that it is Philosophy the one that has to determine what science consists in.

Another thesis comes in hand with the last one; it is a stronger thesis which is logically and scientifically undeniable: Philosophy is science and it is the only sense in which the other disciplines can be called sciences. If their scientificity depends on a philosophical judgment that has to be acknowledged as such; otherwise, to accept such a definition would be an unscientific act. Here one sees how untenable the thought is of those who believe that Hegel's science would be a science of a strange kind different from the other sciences, for these derive their scientificity from the scientificity of Philosophy.

The predictable failure of an *a posteriori* method to define science is acknowledged by Rosenblueth (although without being aware of it) at the end of his essay when he says: "I have to admit that many men of science would not agree with many of the statements that I have made" (op. cit. 89). If scientists do not agree with him it is because accepting Rosenblueth's concept of science would imply that the disciplines they practice are not sciences. This means that Rosenblueth did not include them in the group of treatises he intended to examine in order to see what they have in common; and he did not include them because that would constraint him to an *a priori* concept of science, which could be either true or false, but that is certainly not to be determined by inspecting scientific treatises.

The case of Rosenblueth is completely typical; all the methodologists, without a single exception, do the same thing: Taylor and Wheeler, Margenau, D'Abro, Born, Einstein himself, etcetera. Each of them rules out by means of philosophical arguments (either good or bad ones) a lot of what sciences do, and they bear the hope to find out that what they 'really' have to do is what every methodologist ought to do. If someone raised an objection against this practice or against the exclusion of other elements, the argument would have to be of a philosophical kind.

It is very important for us to underline this point. One only needs to see how Margenau rules out in his discussion of science the necessity to suppose that nature consists in something else than mere empirical and perceptual data.

> Indeed the unsophisticated person who sees no reason to exclude from the nature the tree in front of his house while he himself is away makes tacit metaphysical assumptions which could not stand in the face of Hume's or Berkeley's criticism. Now my personal opinion is that these assumptions are entirely valid and that, therefore, I may just as well include the tree in nature. But I think it important to point out that physics can get along with an interpretation of nature which is not tied to these metaphysical assumptions. (1978, 97).

In the last text of Einstein we have quoted the exact opposite thesis is backed up. This discussion would really have to be philosophical and, of course, Hume and Berkeley would gladly engage in it. If, according to Margenau, physics can do away with what is actually and immediately perceived, the opposite side would state that physics explains with 'forces' many phenomena, even though 'forces' are not empirical data, as Margenau himself recognizes. "Nobody has seen this force of physical attraction, and no one ever will." (ibid., 335). The question whether force is empirically perceived falls within the realm of philosophical epistemology. Even to find out if two statements of physics contradict themselves or not is a logical and philosophical question, which evidently is not resolved by examining *a posteriori* what scientists do. It is particularly philosophical, as we will see in later chapters, the question whether the militant and antimetaphysical concepts they employ depict reality.

Let us mention one more example of this. After listing concepts such as amplitude, probability, crystal, atom, potential, magnetic field, etcetera, Margenau adds: "These elements are not and can never be part of sensed nature; they serve in fact the purpose of explaining nature. By physical explanation the physicist means nothing more than the establishment of organized relations between constituents of nature and these elements. Every other interpretation, as the search for the Causes of Things, draws in more metaphysics than the minimum we are admitting." (op. cit., 98) It becomes clear thereby that the concept of science professed by Margenau is determined by philosophical conceptions. How logic or antiscientific is to admit that 'a little bit of metaphysics does not hurt anyone' is something to be determined by Philosophy. Likewise, to determine up to which point an 'organized relation' really constitutes the explanation of something is a problem acutely philosophical, for it is clear that by an 'organized relation' Margenau understands a law (i.e. a preposition) that says whenever we have x, we have y', but if we ask why it thunders when there are black clouds in the sky and someone answers us 'whenever there are black clouds, it thunders', no explanation is given to us at all.

To put it briefly, the methodologists, who allegedly call themselves scientists, intend to obtain *a posteriori* the concept of science, but first they forget that doing that would be to accept something by authority (an eminently unscientific conduct), and second, they always end up failing in their task. And they could only do so, for without a previous (philosophical) concept of what they look for in the world, they could not conceive the group of scientific things from which they would extract, by means of generalization, the concept they were supposedly searching.

Rosenblueth discovers that explaining is not a task of science. Margenau discovers the opposite. This is an evident proof of the ingenuity of those who believe that the concept of science is attainable by means of induction.

If we stick to what contemporary scientists think, the harvest we collect is extremely contradictory. In each of the following disjunctives underlies two very different definitions of science:

1. Some think that reality is as is it is given to us by empirical data and that science essentially investigates reality. Others think that science deals with phenomena and that it cannot determine whether we know reality as it is.
2. Some scientists hold that it is unscientific to draw suppositions. Others hold that science has to suppose some things and has to start with a group of conjectures.
3. Some say that Mathematics is not a science (cfr. Rosenblueth). Others say that it is.
4. Some say that Logic is not a science. Others assert the opposite.
5. Some affirm that science has first to determine with precision of what it speaks. Others emphatically deny this (cf. Taylor and Wheeler).
6. Some hold that science is essentially experimentation. Others contradict this.
7. Some say that the theory of Darwin is not science. Others, specially the most part of the biologists, believe it to be so.
8. Some believe that it is a task of science to know things. Others judge that the task of science is to predict future events so that we can manipulate phenomena. Here we have a definition of science as knowledge and a definition of science as a technique.
9. Some vehemently affirm that only what is expressed in quantities is scientific. Others strongly disagree with this.
10. Some say that psychoanalysis is not a science. Others say that it is.
11. Some affirm that the task of science is to explain phenomena. Others believe that science's only task is to describe and to draw relations between distinct elements.
12. Some think that historiography is not a science. Others think that it is.
13. Some think that Sociology is not a science. Others disagree. Etcetera.

In addition, of course, there is the vulgar conception according to which science consists in collecting empirical data. Evidently, the man in the street has not realized that this conception implies that Mathematics is not a science, something he would not be willing to defend. Besides, he would not pretend to be smarter than Einstein, who explicitly prevented us against the reliability on our senses:

> The discovery and use of scientific reasoning by Galileo was one of the most important achievements in the history of human thought, and marks the real beginning of physics. This discovery taught us that intuitive conclusions based on immediate observation are not always to be trusted, for they sometimes lead to mistakes. (1984, 14).

As we have seen before, the discovery that Einstein attributes to Galileo was made twenty-two centuries ago by Philosophy and not by physics. But regardless of this, what is important is that the untruthfulness of empirical data imposes itself to anyone who reflects a little upon the matter. Empirical data tells us that the sun 'rises' in the morning, but if the earth is the one that spins around the sun, as it will be clearly admitted nowadays, to say that the sun 'raises' is completely false. Empirical data affirms the existence of the surface of the table, on which I write, but nowadays we know that there are no continuous or tangible things, that there is 10,000 times more void space than space filled with matter, and that matter itself varies according to the movements of its electrons. Surfaces do not exist. Empirical data provides testimony of the stillness of the lamp, but nowadays we know that the lamp and the house in which it is are traveling at a speed rate of 30 km per second. It would be ludicrous and false to reply that we distort empirical data appealing to other empirical data. Such an opponent would demand a justification for our relying on other sensible data since we were previously deceived by them. Evidently, it does not depend on sensible data whether we rely on something, for both things are equally attached to the senses.

As Einstein says, scientific thought has taught us to distrust immediate data. Hegel had already said this when distinguishing skepticism (scientific skepticism) from the half complete skepticism that prevails nowadays: "(True) skepticism was essentially alien to considering the objects of our immediate certainty as true" (GP II 375). This immediate certainty is what some unreflective scientists, following vulgar conceptions, believe to be a magic wand with which they can demonstrate something and interrupt thereby the *processus ad indefinitum* with which one comes across when everything has to be demonstrated: in order to demonstrate we need prepositions, and these prepositions have to be demonstrated as well, etcetera. But sensible certitude does not work magic like that; it does not break the *processus* for the simple reason that it is not trust-worthy. In regard to authentic skepticism,

Hegel says many times that his Philosophy, the true science, is "that skepticism which is fulfilled" (PG 67). It does not accept anything by authority, not even the definition of science, and hence it is more scientific than the disciplines so-called sciences, for each of them take this definition for granted.

Let us keep this in mind: although it may be legitimate for the individual scientist to suppose some things, science as a whole cannot do this, for if it supposes something it does not demonstrate and hence it is not science. "First, we cannot make suppositions; this is a big principle which is extremely important" (GP III 128).

It seems obvious to me that Hegel's science is interpretatively castrated if we understand it as if it were different from the scientificity towards which other sciences aspire or as if it worked on its own kind of basis. Hegel is extremely clear in that regard: "There can only be one method in every science and knowledge. Method is the concept which displays itself, nothing more, and it is only one." (PR I 62).

"On this regard one must briefly say that, despite the others' conceptions of what Philosophy is, I consider the exercise of Philosophy something intrinsically attached to scientificity." (Ästh I 50).

Hegel reproaches Kantian philosophy precisely what Hegel's commentators want him to do: "Kantian philosophy has been unable to have any influence at all on the treatment of sciences. It leaves the categories and the method of ordinary knowledge completely untouched." (EPW 60A)

Those who make Hegel innocuous for sciences banish him from the world. Findlay is an example of this: "There is no doubt that Hegel believes in the complete validity of ordinary scientific methods at the level at which they are applied. His criticisms of them are a matter of 'second review', a consideration of them from a vantage-point foreign to them as scientific." (1958, 349) As we will immediately see, this irenism pretends to place Hegel within the non-agression pact with sciences in which many Philosophy professors have decided to live peacefully and with professional prestige. If we asked Findlay what does he understand by science when he says that Hegel accepts the ordinary scientific methods, the vulgar conception of science would certainly float to the surface (a conception which, as we have already examined, destroys itself with the first act of reflection); or at least, the belief according to which it is possible to determine by means of an *a posteriori* survey what science is would reappear, but we have seen that such belief is a myth.

The same happens with Lauer when he says: "The 'science' of which Hegel speaks bears little resemblance to the contemporary notion of science" (1982, 222). But there is no contemporary notion of science! I mean, aside from the vulgar one that annuls itself. The most learned ones we have mentioned in our list considerably differ from each other and it is impossible to find a common denominator between them.

Findlay and Lauer exemplify the extremes of all Hegelian scholars. The former wants to castrate Hegel because he hates his theses, while the latter, with great sensibility and merit, accepts all the Hegelian theses as a whole, but not scientifically. And that changes everything, even the meaning of the theses themselves. Even *a priori* it is unthinkable that the parts remain unaltered once we take into account the sole purpose of Hegel's works: "The rebirth of all sciences through Philosophy" (JS 170), for a "new epoch has arrived in the world" (GP III 460).

What has happened to scholars is that, having forgotten that "the spirit has left Art behind" (PG 492), they confuse Philosophy with Literature. Let us see what Findlay says: "...his transitions are only necessary and inevitable in the rather indefinite sense in which there is necessity and inevitability in a work of art." (1958, 74). And Laurer too:

> The greatness of the *great* philosophers does not rest on the inner consistency or on the convincing power of such rational elaborations. Rather, their greatness consists (as does that of the great poet or great artist) in the quality of their experience, its capacity to reveal in a new way the possibilities of human experience. (1977, 25s).

The parenthesis is of Lauer himself.

Those who avoid Hegel's scientific assault in defense of what there is today, get involved, voluntarily or involuntarily, with the modern 'syndrome of science', which is the front-cover of half complete skepticism. These skeptics believe blindfoldedly in science in order not to believe in other things (the most important ones), but they do not know what science is or what it says, nor they care to find out; for they do not intend to embrace it; they want to leave it aside so that 'everything else' becomes relative and suits the taste of each person. It is very easy to be a fanatic of science without being obliged to say what one understands by science. As we said, they do this blindfolded; they accept to be told what science is and what is not. "What would the ancient skeptics have said of that illegitimate skepticism which dares to look for reconciliation with the bare dogmatism of these sciences?" (JS 226) This text serves us to evaluate the untruthfulness of some commentators when saying that Hegel believes in the validity of ordinary scientific methods.

That is the syndrome; no one knows what science is; people point their fingers at each other and no one ventures to defend with arguments —that is to say, with Philosophy— his concept of science. In addition, it is not true that one appeals to what others have said, for while examining what others say or do, one only considers to be scientific that which one had in mind before, so this turns out to be a strategy in order not to defend one's own concept of science, which is only negative since its function is to exclude the themes which are truly important.

While comparing Sextus Empiricus scientific skepticism with modern, half complete skepticism, Hegel says:

> The invincible figures of speech of dogmatism were employed by Sextus Empiricus with great success against dogmatism, especially in Physics, a science that, along with applied mathematics, is the true empire (…) of the restricted concepts of the infinite, but physics is for the modern skeptics a science which challenges all reasonable incredulity (JS 246).

Let us point once and for all which is the naive dogmatism of the contemporary sciences that claim to be empirical, for that will be the bridge for our next chapters:

> The fundamental deception in scientific empiricism is always this one: it employs metaphysical categories like matter, force, unity, plurality, universality, infinite, etcetera. In addition, it joyfully makes inferences on the grounds of these categories and in doing that it presupposes and applies the forms of inference, and during this process this empiricism does not know that it contains itself and applies metaphysics, and uses the said categories and junctions in a non-critical and unconscious way. (EPW 38 A).

"As a matter of fact, they cannot do without the concept, but by some sort of tacit arrangement they run certain concepts, such as 'parts' and 'forces', without knowing in the least if they contain a truth or *how* they contain it." (GP II 171)

It is impossible to force Hegel into this non-aggression pact that prevails in university faculties as well as in the whole world, for that would mean to embrace the 'autonomy of sciences', which is preached by the Church and which is only respected by it. Hegel, who is on the same side as the Church, is not willing to observe this pact, since he explicitly says: "The physical atomism (*die Atomistik*) places itself face to face before the idea of a creation and a conservation of the world by a different being. But suddenly the research of nature feels that, on account of atomism, it is exempted of having an explanation to the world." (GP I 361) "In this regard the step contradicts the principle of these sciences [...] On this point the spirit is equal in Catholicism and Protestantism" (WG 913); "the excessively mathematical point of view that identifies the quantity —which is a particular stage of the idea— with the idea itself, that point of view is nothing else than the principle of materialism" (EPW 99z). "Materialism and naturalism are the following system of empiricism" (EPW 60A).

To a non-aggression proposal Hegel answers: "But there are false truces" (EGP 196). "It is a false idea: there can only be *one* truth." (EGP 290)

Neither sciences nor Theology can pretend to achieve autonomy from reason, and we saw in fact that they never achieve independence from Philosophy, for it is always Philosophy or a pseudo Philosophy what decides what is scientific and what is not.

The initiative of Hegel, to reach at last truth scientifically, is the most ambitious enterprise that reason has ever set out on. But it will be accused for being prideful and excessive by him who does not understand that "the internal knowledge of the scientificity of knowledge lies in the nature itself of knowledge" (PG 12). It is an initiative that sooner or later someone had to take. It is not possible to think that human reason would have eternally resigned itself to this irrational languor in which convictions are adopted because one likes them, or because they are useful, or due to any other motive that is not their scientifically proven truth.

Far from being prideful, it is an authentically humble enterprise, as Hegel repeatedly warns: "The rational is the real path in which everybody goes and nobody stands out." (Rph 15 Z)

> Thinking is what is universal: by this means one places oneself in the point of view of the rest of mankind and renounces to the particularity of his own heart and genius. In that moment all peculiarities step aside and are

> cancelled. Thinking is what is universal in and by itself. Living on such grounds is the highest humility. Humility means to make aside every distinction, every peculiarity, and to submerge oneself into the universal. (EGP 282)

"Philosophy is not something particular as an artwork" (GP II 23). Into that position of universality, not of originality, this present book steps forward. For that reason, even though I think like this, even though this is my system, this book presents itself as an interpretation of another philosopher who, on his part, incorporates Plato and Aristotle and all the history of Philosophy. I prove thereby that I actually believe that truth is knowable to the human mind, for I make clear that it has certainly been known before.

The only way to interpret Hegel is to give more weight to demonstrations than to interpretations; otherwise, one does not make Philosophy but literary criticism. Hence our title is *Hegel was right.* There are certain questions, especially those which are naturally linked with the subsequent development of sciences after Hegel's death, in which Hegel only outlined the path which demonstrations ought to follow; the present book cannot leave that only as an outline, which means I will not expose how Hegel thinks without engaging my own views. The only way to do this is to rely only on what is demonstrative, that is to say, on what is truly philosophical.

CHAPTER II

Why the subject?

What we will put forward on chapter three in regard to the ontical consistence of the subject is not only the key to Hegel's philosophy, as he repeatedly warns, but the key to all Philosophy and all sciences, and it is the most important discovery that the human mind has ever made in all history.

However, there is a certain school of thought which obstinately refuses to take the subject into account, and which even says that science should forbid itself from making any consideration on it. Our present chapter examines this resistance and demonstrates that, far from being scientific, this resistance is scientifically untenable because, once we do away with the subject, it becomes impossible to provide scientific concepts with any meaning at all. If we do away with the subject, sciences do not know what they are speaking of. Our present chapter also dispels one of the most widespread and inveterate prejudices there are: the belief in the empirical origin of concepts.

1. Something about modem physics

We must first say that our century has not contributed with any single wonder to this parade of obstinacies which deny the subject. In fact, in order to avoid it, those positions claim to rely only on the physical manifestations and therefore reduce all realities —even the psychic and biological ones— to physical processes, in the hope that this systematic and "objective" approach will scare away the subject once and for all. But Physics itself has proven these hopes barren by reinserting the observing subject into the picture, and this is not something trivial but the absolutely decisive factor that determines the characteristics of any observed objects or phenomena: both relativity and quantum physics make all objective data to depend on the subject. Here we see the subject reappearing where they never expected him to be.

One can barely restrain the laughs by reading Taylor and Wheeler when they say: "*The word 'observer' is a shorthand way of speaking about the whole collection of recording clocks* associated with one inertial frame of reference." (1966, 19).

They fill up every inch of the universe with watches and rods whose masses and fields would distort even the most robust and delicate physical phenomenon, leaving it unrecognizable and unobservable —which was the only thing at stake. Watches and rods without masses and fields are a physical impossibility; consequently, the unperturbed phenomena of which Taylor and Wheeler speak are not *effectively* observable and hence inexistent to modern physics. Taylor and Wheeler have not yet come to understand that, according to the theory of relativity and to quantum physics, any speculation about phenomena that is not really observable falls outside the realm of the physical sciences.

The acknowledged physic C. F. von Weizsäcker sums up very well the present state of affairs of his discipline: "An object is an object for subjects in the world. This would be made even clearer if the concept of object could be reduced to decidable alternatives." (Bastin 1971, 253)

Avoiding the subject is thus reduced to a simple *wishful thinking* that lacks any kind of scientificity.

Leon Brillouin provides a more extended analysis of the situation, but the quote here presented is worthwhile examining; I display it here for the best interest of my readers who are not as well acquainted with modern physics:

> I become very suspicious whenever I hear the word 'given'. There is only one occasion when it has a definite meaning; this is in the statement of a problem given by an examiner to some helpless students. In this situation the velocity is supposed to be exactly the given velocity, with no possible error or

> discussion. But in real life, this never happens. If I observe an unknown moving object in the sky, nobody can give me its velocity. Whether it be a star or a flying saucer, I have to measure the velocity by some experimental device. I can use optical signals, which would be reflected from the unknown object, to measure the delays, the Doppler shifts, etc. From these measurements, I can compute the velocity, but I should always be aware of the fact that these very experiments always perturb the motion. The velocity after observation is not the same as before observation. Every experiment requires some coupling between the observer and the observed object,... (1970, 4).

The photons with which I have to bomb the object to make it visible exert force upon it and modify its state. And I cannot talk of an unperturbed object because "unperturbed" means 'not observed', something which does not exist to Physics.

To modern physics there are no objects without a subject. If someone believed that by relying on Physics one would put the subject away, his disappointment could not be greater.

As we will later see, what Hegel has to say in order to make the study of the subject scientifically unavoidable is ten thousand times deeper than the entire contribution of Physics in our century. It is not a matter of chance that Physics has finally consented to be more reflexive than the physics of Newton. However, it is timely to point out that the negation of the Newtonian absolute movement and space, a negation which is the point of departure of the theory of relativity, was already implicit in Hegel: "In the empty space there is no movement, for there is only movement in relation to a different place." (NH 126).

According to Newton and his master and namesake Burrow, although there is no other object towards which the studied object approaches or from which it goes away, the said object would be moving, for its movement is absolute. It takes place in the absolute space which is the *sensorium Dei*. In other words, although it does not approach to or go away from any other thing, God would see that object moving.

The idea according to which the absolute space is the *sensorium Dei* goes back to Henry More, and in a certain way, to Pierre Gassendi. But we must take into account that in times of Newton everybody believed that movement was something absolute. And they still think the same thing, all those who have neglected relativity in order to remain attached to the myths of common-sense. Since the times of classic physics, Hegel disproved that belief with the mere analysis of the idea of movement:

"Ultimately, it is absolutely clear that movement as such only has meaning and existence in a system of many bodies, and by the way, many bodies which stand in reciprocal relation according to different *determinations*" (EPW 269A).

For the same reason, Hegel rejects the first law of Newton: "a rectilinear infinite movement is a void mental monster; because movement is always towards something" (GP II 193).

Moreover, Hegel holds that the physics of his times would have avoided those phantasmagorias if they had taken the time to read well Aristotle: "Aristotle shows that the void suppresses movement [...] In movement the body —as distinct— is a positive relation, not towards nothing." (GP II 185).

And a little before that point in his book, he says: "Aristotle deals then with the *void space*, an ancient question for which Physics still cannot find a solution. They could do that if they read Aristotle, but for them thought and Aristotle are things which have absolutely no existence in the world."

Already with his commentary on Zeno, Hegel had explained the relativity principle of movement: "And this is also true: that movement is definitively relative." (GP I 315).

In order to reject the notions of absolute space and time as unscientific, Hegel did not have to wait until Michelson's experiment: he only needed to realize that those were mere abstractions. Evidently, an abstraction cannot pretend to have the same status of reality: "The word *absolute* has often no other meaning the word *abstract*; thus, absolute space and absolute time are no other thing that abstract space and abstract time" (EPW 115A).

We have displayed these quotations in order to show that the link between modern physics and Hegel's philosophy is not fortuitous. Our author already knew how unscientific and ungrounded it is in Physics to project as real something not observable.

2. A BASIC PRINCIPLE

Before venturing into the refutation of the different ways out that have been invented to avoid the study of the subject, we beg the reader to consider the following reflection as fundamental, for even though it is obvious in itself, it has been neglected during the last centuries.

If a word does not have an empirical meaning, the origin of the concept in question cannot be sensation, and hence it is necessary to look in the subject itself both for the origin and the meaning of it.

For all what follows in this book, the above reflection is essential.

It cannot be stated that the origin of such a concept is sensation, for the simple reason that the origin would have to be an empirical data, but precisely no empirical data corresponds here to the concept.

For example, the concept of 'point', in which definition enters the idea of unextendedness, could have not been caused evidently by any sensible data or by imagination, for every empirical data is extended and every image of fantasy is extended as well; consequently, the origin and the meaning itself of the concept 'point' should be looked for in the subject.

On the other hand, all sciences use at least some concepts whose meaning is not empirical. The science that raises fewer doubts in this regard is Physics, but apart from the example I just mentioned, we already saw that Margenau listed only some of the few concepts which are not empirical: mass, energy, charge, force, wavelength, strength, potential, probability, amplitude, crystal, magnetic field, atom, photon, electron, meson. (1978, 98)

The following text of Hegel deals with Mathematics, but it mentions several terms employed by Physics as well:

> Other mathematical determinations such as infinite, relations, infinitesimal, factors, potencies, etcetera, have their true concept in Philosophy itself; and it is wrong to believe that Philosophy should extract and take them borrowed from Mathematics, where they are accepted without concept and often without any meaning at all; these concepts must wait until Philosophy gives them their sense and justification. (EPW 259A)

It seems evident to me that none of these terms have an empirical origin.

Einstein, the most acclaimed physic of our century, goes far beyond Margenau. His forceful statement should not be so zealously hidden away from the public by the apologists of the empirical sciences:

> And let us say this in regard of the historical development. Hume clearly understood that certain concepts, for instance, that of causality, cannot be extracted from the material of experience by logical methods. Kant, convinced

> of the indispensable character of certain concepts, considered them —just as they are selected— as necessary premises of any kind of thought and distinguished them from concepts of empirical origin. But I am convinced that such differentiation is mistaken and does not approach the problem in a natural way. From a logical point of view, all concepts, even the closer ones to experience, are free constructions, just as the concept of causality, which was the starting point of the dispute. (Schilpp, I 1969, 12).

According to Einstein, not only some concepts of Physics but every single concept lacks an empirical origin.

3. What everyone understands?

That being said it is now of primal importance to take this into account. More often than not, one believes to 'know' the meaning of a certain word, because one thinks that the meaning is a determined empirical data which is easily observable. But in the cases that the word being considered does not have an empirical meaning, it would be unjustified to suppose, without further investigation, that 'we know what it means'. It is by no means true that we already 'understand' its meaning; we may be *imagining* something, but that image is not the meaning in question, for imagination can only contain combinations of empirical impressions and accumulate the data provided by the senses; however, if the concept does not have an empirical meaning, no empirical impression or combination of them could provide it. Let us go back to the above mentioned example. There are some people who believe they know what the meaning of the word 'point' is; but what they imagine is a speck, not a point; if it were unextended, as it should be to be a point, they could not imagine it. As a matter of fact, they *do not know* what the word point means; they *do not have the concept*.

We must look for the meaning and for the concept in the knowing subject.

Let us put another example. In the context of discussion of modern pluridimensional, non- Euclidean spaces, a great number of mathematicians and physics do easily away with the problem by saying that space is a group of points. However, since they do not know what a point is, this so-called definition lacks any kind of meaning. The situation gets even worse if they try to define point, for they will irremediably say that it is an unextended part of space. They define point in terms of space, and they define space in terms of point. The circularity of this argument is manifest. We do not gain any knowledge by means of this procedure, and clearly, one will notice that these people do not know what they are speaking of.

Here is where what we were discussing occurs: they ease themselves by believing that the meaning of the word space is a certain empirical data that is easily verifiable. An opponent would probably answer to this objection something like 'you are not going to deny that I am looking at the space between my body and the wall in front of me'.

Of course, I will deny this. What this person is *seeing* is the wall in front of him. Space as such is not observable; it is not an empirical data.

Since it is not an empirical data and they do not know how to provide it with meaning by other means, they do not know what they are speaking about.

One brief digression: if modern physics boasts their rejection of non-observable entities, one wonders why they keep speaking of spaces, let alone pluridimensional ones. The only justifiable criterion would be the 'necessary factor to explain observable data'. But we will see in our fifth chapter what the *Science of Logic* has to say about those explicative intended factors.

It is the right time to do away with a false way out. Scientists will admit that many concepts do not have an empirical meaning, and yet they say that they are *inferred* from empirical data. I do not deny here the logical legitimacy of the inference in question; but it is obvious that, in order to infer certain concepts, we must first need to know what they mean. Precisely, in this case, the meaning does not coincide with any empirical data, for it is necessary to infer it since the meaning is not an object of empirical experience; if it was an empirical data, we would not have the need to infer it. In some of the premises of the inference process, one of these concepts —which are not derived from empirical experience— appears suddenly: How is it possible to handle it if we do not know what it means? What it seemed to be a way out only makes more acute the need to look in the knowing subject for the origin and the meaning of concepts.

We do not need here to stop for long in order to examine another way out which is identical to the previous one. According to this position, there are some things that are not empirical in themselves but only *in their effects*. God and the soul also have sensible effects, but that is not enough to know what those terms mean. The question is not about their empirical effects: what is here discussed is if the entity itself to which its causality certain effects are attributed is empirical or not. Although we do not put into question the legitimacy of such attribution, it is impossible to know which entity they are speaking of as long as we do not know the meaning of the word they designate it with. Moreover, to judge if the said entity can be the cause of such effects, we need to understand beforehand the word that designates it. The fact that the said entity could have or not such effects does not make us know what the meaning of the word in question is.

Now, he firstly hinted and then explicitly said that, more often than not, the men of the street or the scientist do not know what they are speaking of; they do not have the concept; they do not understand what they are saying. It sounds severe, but it should not offend anybody because this is not any kind of truth: it is a truth that it is urgent to straighten out despite its harshness. "To know what one says is much more unusual as what one thinks, and one is very unjust by calling severe the reproach of not knowing what one says" (BS 249); "ordinarily, one calls concepts that which is nothing more than a reverie" (WL II 281).

Given that everybody is able to use a certain term not only in private but in a conversation or in a discussion with others, they think that because of that they understand it and that they have its corresponding meaning, when in fact they only have an image product of their fantasy. For instance, as we have previously demonstrated, it seems that common people do not have the concept of 'point'; however, it is obvious that we can all use the word point very skillfully. The same happens with the word space, whose meaning is ignored even by physics.

Let us put an entirely different example: the word 'work'. We all commonly use it without any mayor trouble. Nevertheless, I dare not only all workers but economists and labor lawyers as well, to define 'work'. I have not yet found anyone who has the concept of it. Everybody supposes that they understand this word, but some imagine a certain manual activity (v.g., a man digging a whole with a shovel) and others imagine a still man performing a certain intellectual activity. The most common thing to happen is that each person imagines something related to their own trade or profession.

One only needs a handful of insisting questions to dispel any attempt to define work. Is effort, for instance, something essential for an activity to be called work? On the one hand, if one answers affirmatively, then the really skillful expert who enjoys his own professional efficacy and performs his job without fatigue would not be actually working despite the fact that his productivity is superior to that of the majority; on the other hand, if one answers negatively, the millionaire swimmer, who is in his own private pool trying to beat the *crawl* Olympic records, would be working, —which is an utterly absurd premise.

Does work, in order to be that, need to produce something? We come across here with the physiocrat theses which posit that nature is the only one that produces more than what it plants, and that work only modifies and does not produce things. According to Physics, nothing in fact is ever produced, because the existing matter in the universe does not increase nor decrease. And even leaving this aside, it is obvious that the work of fishermen does not produce anything; they only collect fish from the sea. The same goes for miners and harvesters. Should one person argue that by working one produces exchange value and not something material, that person would confirm in the first place that the majority of people do not know what work is, for evidently, almost the largest part of humankind does not know what exchange value is, and the few who say they do, frequently do not know it. In the second place, it would turn out that a group of workers, who produced a great stock of merchandize which did not find demand in the market and ended up rotting out, would have not worked at all.

Does it lie in the essence of work to modify an object or a material substance? To accept this would be to rule out intellectual work, and that would confirm that one does not have the general concept of work, which was the only thing at stake here. In the second place, and this carries even stronger consequences, the distinction between work and consumption would disappear, for consumption also modifies things. And in the third place, since the fishers, the miners and the harvesters do not modify but only transport matter, it would turn out that the activity they perform is not work. And it would be ridiculous to say that the only thing that matters is the changing of the place in question, for then the thief would be performing a job, with the same right to social remuneration as the workers in the fabrics. In addition, the distinction between work and rest would disappear, for he who goes out to take a walk changes the place where his clothes are in.

The result is here the same as in regard to space and point: he who believes to comprehend and know the meaning of a certain word by using it skillfully deceives himself. What he has is an image "for ordinary life does not have concepts but only imaginations" (WL II 357). And this image is not the meaning of the word in question, because it is too particular and because other people have a completely different image in their heads when they use the same word.

From this follows that the resource employed sometimes by scientists in order to elude the obligation of defining a term is completely mistaken. It is an appeal that is based on what 'everybody understands for…'. But it is false that *everybody* understands. This resource does not lead to the concept nor does it make the scientist in question know what he is talking about.

There would also be a more immediate objection: if everybody understands it, then why don't they say it?

Before examining the next escapist resource it is convenient to state that recent experimental psychology confirms entirely Hegel's observation in the sense that no one should be offended when told that he does not understand and even lacks the concept in question. Let us quote textually the acclaimed author of the treatise *Concept Learning*, psychologist Earl B. Hunt, although his terminology is not the same as Hegel's:

> Conceptual classification may be contrasted with another type of classifying behavior called discrimination learning. In discrimination learning, objects are classified on the basis of directly perceived properties such as physical size or shape. The usual explanation for discrimination learning is that the sensory features of any stimulus are matched to what is already remembered of these features, and that the learner´s response becomes associated with them. The response thus classifies the stimulus. In discrimination learning, subjective representations of immediate and past stimuli seem directly to indicate concrete, physical features (in contrast to the more abstract nature of concept formation). (EB 22, 897, 2).

> While human beings popularly are called abstract thinkers, many of the classifications people make clearly seem to be concrete discriminations. Indeed, people may use the same term either in a discriminative or conceptual way. A child uses the term policeman in discriminating a man in distinctive uniform, while a lawyer may have a concept of a civil servant charged with enforcing criminal codes. (EB 22, 898, 1)

In this example, the professional studies of the lawyer would have lead to the acquisition of the concept, while the child, a typical example of what Hegel calls 'ordinary life', proceeds by means of this association of empirical features called imagination: but the child and the lawyer employ the same word and use it skillfully. It becomes obvious that using a word skillfully does not imply by any means that we have the belonging concept and that we understand. Let us repeat this with Hegel: "To know what one says is much rarer than what is thought" (BS 249).

What Hunt has summarized for us is the result of a large series of experiments. Nevertheless, it is evident that animals also learn how to discriminate and yet they do not have concepts. If we distinguish between science and technique, between knowledge and skills, it becomes obvious that mere empirical discrimination is enough for animals to possess a series of techniques which in many cases are more refined than ours. As we will later see, the technical success boasted by our empirical sciences does not demonstrate by any means that these disciplines *understand*.

4. Leaving terms undefined?

It is likely that the writers of treatises who leave some terms undefined refer to what "everybody understands". It is likely because, if they were not referring to this, their readers and hearers would have their mind in blank after reading or hearing those terms, they would not have anything to hold on to, and they would not know how to cling on to the signals they receive. Although the scientist does not make his reference explicit, his audience notices the reference to what 'everybody understands'. However, since they do not have the concept, the words are not actually referring to anything.

Mario Bunge claims that undefined terms "are not therefore obscure or undeterminate, since they are specified by a number of formulas" (1973, 9).

But this is clearly untenable, for the signs in a formula are even more undetermined when no one points out, by other means, what does each sign stands for, and thus the entire problem reappears out of the blue. By saying what do each sign stand for, we would not employ again undefined terms, unless we want to incur in another manifest, vicious circle. Mario Bunge himself later recognizes that: "But mathematical form alone won't tell us anything about the physical meaning of the formula". (1973, 30)

C.W. Klimster raises another good objection against the procedure of leaving terms undefined, although he says 'concept' where the more appropriate thing to say would be 'word' or 'term': "I cannot accept this idea of understanding because it gives no assurance that we shall know how to apply the concepts in a new situation which may develop in the future" (Bastin,1971, 124). To put if briefly, I can carry out operations the moment I am given a mathematical formula, but if I do not know what its signs stand for, I will be stunned the moment I meet a new situation; I will not know how to apply it. Stripped from meanings, the formula does not hold any relation at all with reality and it is impossible for me to refer it to something.

Mathematics are conceptless; Hegel told us that determinations "were accepted without concept and often without any sense" (EPW, 259 A). This is the reason why I can execute operations like a machine when I am given a formula. Hegel states that with mathematics, the mind finds itself "in one activity that is the extreme alienation of itself, in the violent acti-vity of moving within the lack of thought" (WL I 208).

Those who say that they leave terms undefined, return sooner or later to what 'everybody understands for', but they do this tacitly. It happens here what Bunge says: "A physical theory is assigned a literal and objective interpretation by assigning every one of its referential primitive symbols a physical object —entity, property, relation or event— rather than a mental picture or a human operation" (1973, 119).

The intention to dispel away the subject is shameless. Bunge supposes that everybody understands what words such as object, entity, property, relation and event mean. But all people are far from understanding such things, and even if they did understand them in some cases, that can only occur in function of the subjects, as we will later see: "Giving names, that is easy, but a different thing is to understand" (GP I 241).

We will now raise three definitive objections against the resort of leaving terms undefined, which, unfortunately, is widespread among the rebellious, empirical sciences: they cannot know whether there is circularity and hence absolute vacuity and pettiness in their theory; they cannot know whether there is a contradiction or if they are making metaphysics instead of empiricism.

Bunge himself foresees the perils of this abyss when he demands the 'primitive independence' as the absolutely indispensable requirement for the axiomatization of a scientific system: "the basic concepts of an axiomatic system must be mutually independent, i.e. they must not be defined by others (In fact, if one of them were defined in terms of other basic concepts, it would not be a primitive concept)" (1973, 165).

It is amazing that Bunge closes his eyes before the evident fact that we would not know which, among two terms, is dependent and which one is derived and which one is logically previous if we do not know what each of them mean, that is to say, if we do not define them, which means to give up the farce of leaving terms undefined.

This is a devastating objection not only against Bunge but against all the alleged authors of the method of undefined terms. It is impossible to find out if we are defining A in function of B and B in function of A as long as we do now know the meaning of A and B, or in other words, as long as we do not define them.

It is impossible to exclude circularity if we do not make definitions; it is impossible to avoid that an entire system becomes a big tautology, a bunch of things that that does not affirm or deny any single thing. And this concerns not only scientists who want to leave terms undefined, but also those who explicitly refer to what 'everybody understands'; for if we do not make the meanings explicit, it is perfectly possible that what everybody understands by A is in function of what everybody understands by B and vice versa.

And there is something worse than circularity: contradictions. It is impossible to know if two terms are or not contradictory unless we know what they mean. Most of the times, what people feel tends to be metaphysical, and the only way to prevent that from happening in a particular case is, in fact, to look for the meaning of the term in question i.e. to define it.

According to acclaimed anthropologist Benjamin Lee Whorf, the native language is for the mind a metaphysical frame. Gary J. Millers summarizes thus this position:

> According to this view, the language structure operates as a ready-made metaphysical framework through which the speakers perceive and conceptualize. A person's language is his means of interpreting reality; it shapes his comprehension of the environment and supplies him with his definitions, categories, concepts, and perspectives. (Rossi *et al.*, 1977, 197).

We do not need to embrace another deterministic thesis of Whorf which posits that the preexisting scheme of language is unavoidable. It is possible to elude consciously the prejudices inherent to one's own native language. But precisely, for that purpose, we need to find out what the meaning of the concept is. For instance, if we set ourselves the goal of constructing an empirical science, the only way to investigate if a word is metaphysical and needs to be excluded from the system is to make its meaning explicit. To leave a term undefined is almost as setting oneself to construct a metaphysical system.

5. Definitions by decree

One used frequently to believe that each one of us can define his terms as he pleases, but this ruse has become today very fashionable among certain scientists in order to elude the obligation of investigating the meaning of words, an investigation which, as they can foretell, would necessarily lead them to the study of the knowledge of the subject. We are dealing here with definitions by decree, arbitrary definitions. One simply decides to understand by 'time' such thing and gets rid of trouble.

Let us get to the root of the problem. Every attempt to define tries to lead the comprehension of someone else, to the content which the defining person affirms to be the meaning of the word in question. That is what logicians call the *definiendum*; it would be what we called 'such thing' in our previous paragraph.

There are three possibilities in the arbitrary definitions: that the *definiendum* is another word, an empirical data or a fantastical image or object.

An example of the first modality is the one that we mentioned before: I choose to understand by space a group of points and I get rid of the problem. Such a definition is null and void because point is defined in function of space, and the circularity does not bring any content whatsoever to the understanding of other people. Those who define by decree, obviously, will look as a *definiendum* a word that does not need to be defined in function of the term they are trying to define; but here lies the difficulty of this principle, in order to avoid circularity one needs two definitions which are not obtained by decree. As long as we do not have the true definition of A and the true definition of B, it is impossible to know if A is defined in function of B and vice versa. The only way to prevent the arbitrary definition from being null and void is to abandon the arbitrary definition and use a true definition. This objection settles once and for all the issue.

But there is even more to this. In this first modality, what the arbitrary definition does is simply to affirm the synonymy of two words; nevertheless, what is simply affirmed can also be simply denied. In order to make science we would have to demonstrate the synonymy. The only way, however, to *demonstrate* that two words are synonyms is to provide the true definition of each of them, not an arbitrary one. And here we come across with the same result: the arbitrary definition needs to be abandoned. I mean, of course, if we are really willing to make science.

We must notice that the solipsistic approach does not avoid this difficulty. He who defines by decree cannot be so obstinate as to say that he does not care about other human beings, and that the *definiendum* is *to him* such thing and period. Science means universality, something that is valid always and not only in a solitary act. Since he who affirms a synonymy performs a public deed, he needs to demonstrate his affirmation; otherwise, it can be denied without further trouble. Since all the consequences of the system would depend on the original definition, all the system would depend on the whim of someone who affirms or denies something. If they call that science, I am sorry to say this, but they have confused science with literature.

As we said in the first chapter, science *demonstrates* and Literature only *affirms* and leaves its truth to the taste of whoever wants to embrace it. In that case what one affirms is a synonymy, and both such pseudo-defintion and the entire system are subject to the aesthetical or pragmatic whim of the author, its audience or anybody else. That is a literary essay. The only way to avoid it is to demonstrate the synonymy, but the only way to demonstrate a synonymy is to show the true definition of the two terms in question. Let us say here loud and clear that arbitrary definitions avoid the necessity of studying the subject in order to avoid the true meaning of words, but one deliberately renounces thus to science and starts making literature.

And their resource would backfire at them if they assumed the romantic position of saying things like synonymy is only hypothetical, everything is subject to further revisions, science never reaches an end, the last ideals are always far beyond our reach, and more stuff of that pathetic sort. On a general basis, this position is self-destructive, since there must be something true for a thing to be hypothetical; only in comparison to something else a thing can be labeled as such. But with the case we are dealing here, the hypothetical resort is particularly counterproductive, since 'hypothetical' is not something that can be either true or false, but something which can be either true or false *and which is submitted to a test* in order to see if it is true or false; it is impossible to submit something to a test if we do not *understand* what is going to be tested, i.e. if we do not *define* it with true definitions. The only way to find out if an apparently hypothetical synonymy is true or false is to compare the true definitions of the apparently synonymical words. This is the reason why this resort backfires at them; when a definition is hypothetical it necessarily demands that one looks for a true definition. What can be either true or false and is not tested is called uncertain, not hypothetical. Escapism reaches its utmost point by rejecting the subject and grounding science on sheer uncertainties. One gives up thus science and true knowledge. Let us not make a big deal about this: to give up science means to avoid the study of the subject.

Arbitrary definitions, in their first modality, stumble upon another obvious difficulty. He who advances a term while defining another evidently needs to suppose that such term has a meaning by itself which is not susceptible of arbitrary definitions; otherwise, we would embark upon an indefinite process and, in short, no content would be brought to the understanding of other people. Arbitrary definitions essentially suppose other definitions that are not so. Therefore, since imaginary and empirical definienda are ruled out in this modality, one has to look sooner or later for the meaning of the appointed terms in the knowing subject himself. The same arbitrary definition refers sooner or later to the subject, provided that one does not want to embark upon an indefinite process. As Hegel says: "supposing and determining never reach the final goal" (JS 27).

"It has been made clear that indefinite progress belongs to the reflection which lacks concept; the absolute method, which has the concept as its soul and content, cannot lead us there" (WL II 500s).

The second modality or possibility of the arbitrary definition is that the *definiendum* is an empirical object, a data effectively perceived as empirical by him who makes the definition. If the said person had the concept in question, he would simply give us a true definition and would not come up with the folly of making a dogma out of an arbitrary definition; but since he does not have it, he thinks that he can point out something with his finger and say 'I understand by *definiendum this*'. As we can see, 'this' is, evidently, too particular; which is something that goes against his own interests. For instance, if he says 'by animal I understand *this*', he does not mean thereby that being an animal is to be identical with his dog. It would follow from this that there would not be any other animals, for identity is always reciprocal. If being an animal is to be identical with *this*, other animals would not be animals because they are not *this*.

Aside from the question as whether or not individuality is an empirical data, when someone points out with his finger and says 'this', he aims at individuality. Therefore, this is not what the dogmatic person means to say; he aims at something more universal. He does not have anywhere to land his 'this' other than his own imagination, since he lacks the concept, which would be indeed universal, and the perceived data is too particular. Therefore, the most recurrent modality of arbitrary definitions is the third one.

"It is common in the empirical sciences to analyze what one discovers in imagination" (Rph 32 Z).

The comment of Hegel regarding the third modality is this: "But to remain in the phenomenon or in that which is produced in imagination for the ordinary cognition means to renounce both to the concept and to Philosophy" (WL II 435).

The image presented as a *definiedum* candidate is blurry most of the times in order to avoid being so particular and incurring in the same failure of the second modality. But it does not succeed. For instance, the man of the street, who does not have the concept of animal —even biologists still dispute the definition of animal—, may be blurrily imagining a horse when he says that word. As blurry as it may be, however, it is a horse; should the meaning of animal were that one, it would follow that neither birds, fishes, microbes nor reptiles would be animals.

That is the failure of *every* arbitrary definition, for we saw that, if the *definiendum* is an empirically perceived entity, it becomes even more particular, and if it is a word, it does not actually define anything; it ends up alluding to true definitions which cannot be drawn without the study of the subject. Those are the only three possibilities there are.

Let us provide another example. As we saw in our last chapter, one does not ordinarily have the concept of human. Those who believe, therefore, that, instead of the concept, they have all singular persons before their eyes, only deceive themselves: they are only *imagining* something. A nebulous and vague image comes up to their minds. As vague as it may be, however, it is always too particular. They may be imagining a masculine figure, but that would rule out half of humanity from being human. If they imagine a thin person, that would exclude all fat individuals. Should they imagine a young man, all mature people would be discounted, etcetera.

In this regard scientists do not differ much from ordinary men. The physic who, avoiding the concept, pretends to define space as a group of points, probably imagines a certain group of black specks (not points) against a white canvas, for example, he may be imagining twenty-eight points nicely distributed, but we would rule out thereby the spaces that consist in a million points and the spaces whose twenty-eight points are distributed differently. And so it happens with everything.

One must notice that operationalism always falls in one of the three modalities of arbitrary definitions and fails, and that meditionism is also a mere species of operationalism. Although some specialists believe to have the concept in question, it is obvious that they do not have it and that they are providing as *definiendum* a word, a fantasy or a sensible data; they become thereby targets of the objections we have raised. Let us put two examples: movement and time.

The arbitrary decree of operationalism about movement is the same to say as 'I understand by movement the fact that a body goes from one place to another'. They pretend to operationalize movement, since by means of Cartesian coordinates it is possible to identify and measure the two alluded spaces.

Before criticizing these definitions by decree, let us document its recurrent presence in the statements of contemporary physics. R. A. Serway says the following: "The movement of a particle is completely known if one knows its position in space at every moment" (1985, 28).

Let us give the floor also to Max Born, who, by the way, is very convinced of having the concept: "It is first necessary to subject the concept of motion itself to analysis. The exact mathematical description of the motion of a point consists of specifying at what place relative to the previously selected coordinate system the point is situated from moment to moment." (1962, 16)

And finally, Arthur Eddington: "Motion is generally recognised by the disappearance of a particle at one point of space and the appearance of an apparently identical particle at a neighbouring point."

It is perfectly evident that movement does not consist in that. Although they say 'I understand by this…' they are not *understanding* anything; they do not have the concept. Therefore, they are providing as *definiendum* a verbal expression, a supposed empirical data, or a fantasy. If a particle no longer existed in moment A and then begun to exist again in moment B, there has been no movement or trajectory at all; the particle has not *gone through* the intermediate space; it is false to say that by knowing two positions one knows the movement. By measuring time and using coordinates, physics may determine two points in space and two moments in time, but they can only *infer* the existence of movement as such, and that will occur when they *add* the adequate premises in which the word movement appears for the first time, premises that are not only very dubious but completely gratuitous as well. The idea of movement is *added* from the outside; the operationalist device does not get a grip of movement; at the most, it gets only its effect: the new position of the particle. One of the completely gratuitous premises advanced thereby is that we are dealing with the *same* particle, something of which no empirical observation can bear witness, since identity is not an empirical data but a metaphysical and speculative lucubration without parallel. Another premise is that the intermediate space is continuous, an untenable thesis as our third chapter will show. Another premise is that a new position of the particle can only *explain* if there has been any movement, but should that be the case, movement would be a mere explicative entity introduced from the outside by the speculative mind in order to provide an account of empirical phenomena. If the mind introduces something whose meaning is not the empirical phenomenon itself, then it must define what it introduces so that we understand it and the mind knows what it is speaking about; therefore, every operationalistic definition by decree is accessory, for they need to define movement: the spirit reappears again as though the operationalists had not said a single word.

In the famous 'tunnel effect' all physics agree (cf. EB, 23 717, 2) that it is impossible that the particle has been in, or went through, the intermediate space, for the particle would have there negative synergic energy and imaginary speed, which is simply absurd. It simply occurs that the particle was first in a Y place and later on in a Z place, but there has not been any movement. To affirm the contrary would not only contradict the fundamental principle of modern physics which states that nothing unobservable exists; it would be to affirm a physical impossibility.

The definition by decree of Serway, Born and Eddington and all physics do not get at all a grip of the concept of movement. They provide us as *definiendum* a word or a group of words, or a fantastic image.

But we have shown that no definition by decree achieves its aim of giving meaning to words without studying the knowing subject.

Another famous case is the decretory ingenuity of Einstein when he says: "one understands by 'time' the position of the minute hand of a watch in the most immediate proximity of an event" (1971, 40).

It is obvious that there is no concept here, for sooner or later one would have to define watch, and one would evidently say that it is a machine to measure time, or in either case, one would say something else in which the word time reappears, and thereby we would be immerged in an absolute circularity. At the end of the day, nothing would have been defined; there would not be any concepts whatsoever. Bridgman even mocks that circularity by suggesting that a watch should be defined as a machine that adapts to Einstein's laws. One wonders how serious did Einstein take his own apparently empirical definition when one reads this letter of him quoted by Dyson Freeman: "People like us, who believe in physics, know that the distinction between past, present and future is only a stubbornly persistent illusion." (Dyson 1979, 193).

There is no doubt that other physical definitions of time are circular. For example, some physics say that one only needs to define 'day', which they believe to be something very concrete. They affirm thus that a day is the lapse of time that goes between two successive sunrises. But they cannot define lapse or the concept of 'successiveness' without mentioning time. If there is something which is clearly impossible to be defined without reference to the subject, that thing is time, as we will later see.

If someone says 'I take by time *this*' what happens is that he remains with *this* and forgets about time. This is a valid objection against all arbitrary definitions and exhibits them once and for all as illusions, since no knowledge is possible if the data designated by the expression 'this' is actually time as long as we do not know what the word time means, that is to say, as long as we do not have a true definition.

In other words, I cannot understand someone when he says 'I take by time *this*'. I can understand *this*, but not time. That person does not understand what he/she is saying. In order to understand his/her phrase we would need to know what time is. But then the arbitrary definition would be unnecessary. The arbitrary definition is an *essentially* unintelligible locution. It does not constitute a public act because it is essentially unintelligible. It is a subjective and whimsical act. It falls outside the realm of science.

6. THE PREJUDICE OF ABSTRACTION

There is a more widespread ruse than arbitrary definitions (and which is even more traditional) employed in order to avoid the study of the object. It is the superstition Locke and the Scholastic Philosophers called the theory of abstraction, according to which concepts have their origin in sensation. Few errors in the history of thought are easier to refute than this one; however, it remains attached to the minds of those who do not dare to make a simple reflection. What is striking is that Plato crushed that belief, once and for all, more than twenty-five centuries ago:

"Why, on what lines will you look, Socrates, for a thing of whose nature you know nothing at all? Pray, what sort of thing, amongst those that you know not, will you treat us to as the object of your search? Or even supposing, at the best, that you hit upon it, how will you know it is the thing if you did not know? (Meno 80 D)

Hegel says that thus: to make a generalization "it is indispensable to know what it deals with." (NH 259)

In other words, in order to make a generalization one needs beforehand the concept; in order to abstract the aspect in question one needs beforehand the idea that allows us to focus on that aspect. But then, the concept does not have its origin in sensation; it existed beforehand in the mind because it is the condition of possibility of all the so-called process that leads to the abstraction of the concept from the empirical data.

"To hit the mark with the substantial, one needs to carry along the knowledge of it. Sensation does not know what the substantial is, just as the hand does not (know) what color is." (VG 259).

The object has one thousand characteristics and aspects. Even the person with the most recalcitrant prejudices must recognize that not everyone have in their minds a *determined concept;* but how could someone focus his attention on the aspect that is indeed pertinent and leave aside the irrelevant ones, so that it produces in him the concept in question, if he does not know beforehand what it is about? And if he knows it, he already has the concept; the causal contribution of sensations comes in late. What could one look for if one does not know that?

That does not mean that ideas are innate, as if children were conscious of them from the beginning. The spirit needs to develop and take gradually conscious of itself; concepts only stem from self-consciousness. "The exterior is only an impulse for the spirit to display itself" (GP I 471).

> The spirit, of course, is in itself undetermined, it is the concept that exists for itself; its development is to reach awareness. The determinations that it produces and takes from itself cannot be called innate. This development must be provoked by something exterior; the initial activity of the spirit is reaction; only thus it acquires knowledge of its own being. (GP III 211).

When that superstition says concept, it imagines an empirical impression (v.g. green), not a concept. For that reason they think it is obvious that that impressions could not be caused by anything else than sensations. But concepts are not that. A spatial figure is not a concept. The entire theory of abstraction is based on a confusion of concepts with the images of fantasy. In order to dissuade their defenders, one would only need to remind them, for instance, that the concept of triangle is a completely different thing than the image of triangle: the image is always very much particular; sometimes we imagine our triangle as a white line against a green screen, sometimes as a black line against a white screen, sometimes we imagine it bigger and sometimes smaller; sometimes we imagine it as an equilateral triangle with its vertex pointing upwards, sometimes as a triangle rectangle with a cathetus over its vertical. On the other hand, the concept of triangle is always the same and it is universal.

One of the essential aims of our next chapter will be to show how concepts are shaped by self-consciousness; the task of our present chapter is to make the reader see how useless are all the efforts made to avoid the study of the subject, one of which is the theory of abstraction. Let us see how Hegel reformulates our central argument against it:

> When the unthinking consciousness declares observation and experience to be the source of truth, what it says may well sound as if only tasting, smelling, feeling, hearing, and seeing were involved. It forgets, in the zeal with which it recommends tasting, smelling etc., to say that it has no less essentially determined the object of this sensuous apprenhension, and this determination is at least as valid for it as the sensuous apprehension. [...] What is perceived should at least have the significance of a *universal*, not of a *sensuous particular*. (PG 185)

Had they not determined something, they would not know where to draw their attention. That determination is a concept, and a concept is universal; "only the knowledge of the universal points of view guides to what one must essentially observe" (NH 259). The universal is not caused by sensation or observation; it is the condition of possibility of observation itself: "The observation and experiments, when they are well done, come to conclude that only the concept is objective" (GP III 84).

The theory of abstraction presupposes that the mind picks out one data from our sensations, that is to say, it chooses and selects one of them; otherwise, sensation itself would not be the cause of the concept. For that to be true, however, one of the empirical data would have to be universal. But it is absurd to hold that one empirical data is universal.

If the word abstraction, to abstract, is not only a magic word whose spell makes universals appear, then its meaning must be to extract something that was there before. But evidently there was no universal among sensible data; therefore, it is impossible to say that our sensations produce the universal.

And if, when they meet this difficulty, they leave aside all the etymologies and they come up with the idea that abstraction is a process of generalization by means of which the universal is produced, the difficulty is now of a triple kind. First, one cannot obtain a given concept from any object from the world; the mind would need to focus only on the pertinent ones; but in order to do this it needs beforehand the concept that guides it to them, and this is the reason why the concept is not a product of a generalization but the very condition of it. Second, our central argument has already mentioned this: not all the aspects of the object are essential, and in order to focus our attention only on the pertinent ones, our mind would need to have beforehand the concept that leads to them. Third, if the mental operation called generalization provides something that was not in the empirical data, then that is something provided by the concept, which means that sensation was not the cause of it. If the mental operation does not provide anything, the sensible data would be as particular as they were before, and we would still lack the universal concept.

That is why we called it superstition. The theory of abstraction remains always in a cloud of vagueness; it does not tell us how universals are formed; it believes that by saying 'abstraction!' a universal appears where it was not before.

It is inadmissible that they prop up their theory with the authority of Aristotle, for he explicitly states that we cannot "acquire knowledge without the universal" (Met XIII 1086b 6), which is by the way the same thesis of the *Meno*, but expressed in a less vivid way.

When Aristotle says that "every idea and knowledge is from the universals" (Met XI 1059b 27) he says the same thing expressed in the three last Hegelian texts we quoted.

The same goes to this passage; "When the particular shows itself to him, the knower knows it through the universal" (Phys VII 247b 7): in the immediate lines before, Aristotle denies that intelligence can be modified by external factors and that intellectual states can be produced by those means.

We would strongly like to draw the attention of the reader to the persistent illusion that underlies both these subjects and the more or less diffuse, vulgar epistemology. One believes that if the reality in front of oneself is exactly as described by a certain concept, one can posit that exterior reality is what produces the concept in the knowing subject. This is some sort of magic trick. One believes that, if the real object is simple, that is enough for us to say that the idea of simplicity appears in the subject. If the object is identical with itself, this is enough to explain the production in the mind of the subject of the concept of identity. If the object is contemporary, that is enough to explain the formation in the subject of the concept of contemporaneity.

By looking at these concepts the reader realizes that such belief is untenable. We can convince a completely illiterate man that the object in front of him is identical with itself, but that is not something that he thought of; he did not had developed the idea of identity in spite of having in front of him an object identical with itself. Therefore, the cause of the said idea is not its belonging reality. If we clearly explain to that individual what substance is, he will agree with us that the objects in his surroundings have always been substances, but he would have never come up with such a round-bout, metaphysical consideration; he did not have the concept despite that he had all the time the reality belonging to it before his eyes.

Although this persistent illusion is unsettlingly crude, many scientists fall in its web when they sing the praises of 'mere observation': they really believe that one only needs to open his eyes so that the belonging concepts appear in the mind of the knowing subject. On the same rough illusion the theory of abstraction we refuted is implicitly based. Its authors were convinced at first that exterior reality is the cause of concepts, and since exterior reality is empirical, it turns out that sensation is the cause of concepts. How exactly, is the causal process is a question they think to be secondary; that is why they leave over it a cloud of vagueness.

We do not need to stop for long in a resort to scare away the subject that is widespread among positivists. It states that it is not necessary to find out the meaning of words because the meaning of each of them is the sum of all the objects to which it is applied. In the first place, they would have to indicate what does the word 'all' means, an impossible task without taking the subject into account. Thereby, this way out is thwarted. But most of all, since they do not have before their eyes or in their imagination all the objects that the word in question predicates, the only way to refer to them is to understand the meaning of this word. The defenders of this way out do not know what they are talking about, neither do we know to which objects they refer. Hence the meaning is needed beforehand, since it is the condition of possibility of the way out itself. The meaning rather reduces this resort to the third modality of decretory definitions; the one that naively believes to have the *definiendum* in the imagination. But that has already been refuted.

7. Coup d'état against the subject?

Let us get finally to the big ruse that intends to scare the subject away. It is some kind of coup d'état on the part of many scientists and can be formulated in the following way: we base ourselves on the objects, the real, the material, and what is empirically observable; science is not interested in the subject.

Hegel answers: "I have been only too often and too vehemently attacked by opponents who were incapable of making the simple reflection that their opinions and objections contain categories which are presuppositions and which they themselves need to be criticized first before they are employed." (WL I 20)

Everything leads to think that the authors of the aforementioned *coup d'etát* have not realized that 'object', 'material', 'real', etcetera, are categories and concepts; they have not realized that, if they do not define the words, not even they will know what they are talking about and neither will we: none of them is definable if we do away with the subject.

"Physics do not know that they think like that Englishman who was happy because he knew how to speak prose" (GP III 426).

> In this Newton is such a perfect barbarian with concepts, that it happened the same to him as what happened to one of his countrymen who was completely surprised when he knew that all his life he had spoken in prose, for he had not realized that he was so smart —only that Newton did never know, did never realize that he had concepts and that he had to deal with them, by believing he had to deal with physical things. (GP III 231).

The reality that lies before us, as we said, has a thousand characteristics and a thousand aspects. It is impossible that the scientist or his addressee refer to a particular aspect or characteristic if they do not consider the concept of that reality and its defined meaning. Sign language is of little use here. To point something with our finger has a thousand meanings: Does pointing out something means that the object is present? Does it mean that it is visible? Does it mean that it is real? Does one intend to point out the place where the object is located? Does doing that means that it is modern? Does it mean the exterior? Does it mean the interior? Does it mean that it is contemporary? Does it mean that I like it? Does it mean it has this color? Does it mean it has this consistency? Does it mean it is 'like this' and not that it is 'this'? Does it refer to the genre of the present object or to its species? Does it mean it is as 'healthy or as 'faulty' as this thing is? The questions are endless.

"The thing cannot be different to us than those concepts which we have of it" (WL I 15).

"It happens as always that one alludes or mentions a perception or an experience; as soon as man speaks there is a concept there, there is no way of making it aside, it reappears in the mind as a clear sign of universality and truth, since it is precisely the essential" (GP I 336).

> Without thinking and concept, the object is only an imagination or also a name; it is only in mental and conceptual determinations that it *is* what it is. In fact, those are the only ones that matter; they are the true object and content of reason; and what one understand as object and contrast by comparison has only meaning by them and in them (WL II 493s).

Only science, in its strict and rigorous sense, can do this. The unmatchable contribution of Hegel has two initial steps that define everything. First, it states that scientists are dealing with concepts and that they cannot do away with them. Second, it poses a very simple question: Gentlemen, these concepts that you have found 'there', where do they come from? Or more importantly, how do we know if they have a meaning? How do we know if they capture reality?

It is useless to answer by pointing out some data of the environment, for we cannot know whether that data corresponds to the concept in question so far we do not know what the concept means. Decretory definitions have been already ruled out.

"By the way, they cannot do without the concept; but by some sort of tacit agreement they let certain concepts run, like 'parts', 'forces', etcetera, without knowing in the least if they contain some truth and *how* they contain it" (GP II 171).

We beg the reader to pay special attention to this passage:

"In the modern, analytical form of mechanics such propositions are put forth simply as results of the calculus, without enquiry whether by themselves and in themselves they have a *real significance*, i.e. one to which there is a corresponding physical existence and whether such meaning can be demonstrated." (WL I 276)

Let us first notice that this text echoes Aristotle. Having pointed out that science borrows terms from ordinary language, Aristotle says "Everything that is intelligible and primary to most people is often barely intelligible and contains little or nothing of reality". (Met VII 1029b 3).

But Hegel goes way beyond that, for he holds that the knowing subject must necessarily look for the meanings within himself. What has been put forth in this chapter gives him the reason: "Science cannot start by something objective, but by what is not objective which makes itself into an object for itself as 'original duplicity'" (GP III 430). This last expression he borrows from Schelling.

The above quoted text WL I 276 rubs salt into the wound: scientists demonstrate and prove (or believe that they are proving) many things; but if the preposition that expresses what is demonstrated or proved is not understood, if we do not know what the meaning of the words used in that preposition is, then not even scientists themselves know what they have demonstrated or proved. It is as if they had not demonstrated or proved anything. "It has been demonstrated but it has not been understood" (GP II 591).

In every demonstration concepts have to be employed. Therefore, the vigor and strength of the demonstration depends definitely on what the concepts mean. The contribution of Hegel consists first and foremost in the sole method by means of which it is possible to look for the meaning of concepts. That is why he called *The Science of Logic* his most fundamental work. It has been an astonishing superficiality to believe that he deals there with things that do not have to do with logic. Logic consists in 'what follows', but what determines if a preposition 'follows' from other(s) is not exactly the meaning of concepts.

As far as proofs go, let us say the following: it is useless that scientists *show* with emphasis an empirically perceived fact; they do not confirm or demonstrate thereby the preposition they are defining since we do not know what it means. It is pointless that they affirm, once and again, that the alluded empirical data corroborates what the said preposition states, for we do not know whether the meaning of the terms employed therein corresponds to the said empirical data. First, one needs to find out what do these terms mean *by themselves*; only thus we can compare that meaning with the empirical data and see if they match.

Let us suppose for a moment that the technological success which is recklessly boasted against the 'ineffectiveness' of Philosophy demonstrates a thesis of the scientists. The problem is no other than this: What is here demonstrated? What does the so-called demonstrated preposition mean?

If we do not understand that preposition, what do we obtain when it is demonstrated? It has not even been proved that the said preposition has a meaning.

He, who alleges that the technical success of physics demonstrates the reality of the concepts they use, would have not understood anything. The smith and the carpenter, for instance, relentlessly use the idea of surface, they work with surfaces and they obtain certifiable successes day after day; nevertheless, surfaces do not exist: there is not even one single surface in the entire world. If logic is due any respect, this sole example would suffice to show that successful practical results do not prove that the ideas which inhabit the mind of the subject correspond to reality.

Almost all the technology of the prehistoric man was based on mistaken assumptions, and yet it had sensational successes, which are in fact more important than those of modern technology. Among the successes of the former we find the wheel, the plough, the slingshot, the knife, the hybridization of seeds, navigation, fire and agriculture. I do not stop here to give a detailed account of the mythological ideas which lead the prehistoric man to those monumental inventions; let us only say that the aqueduct was invented by men who held the belief that the Nile was a god.

Moreover, concepts are not even needed for practical efficiency, like we see in the case of animals, whose technique is often much more refined and skillful than ours. Our aeronautics makes an effort that is still clumsy and rigid to imitate the wings of birds, their plasticity and their adaptability to different circumstances and purposes. Man has not yet invented such delicate radar as that of bats and many fishes, or such heat-detecting mechanism as that of reptiles. If one does not need theories to achieve efficiency and practical success, there can be efficacy with mistaken theories, for the result does not depend upon them. The same goes to tautological theories with no grip of reality: this is the most recurrent case in physics. A tautological theory equals to no theory at all. No theory at all is what animals have. As we have mentioned before, it has not even been proved that the said preposition has a meaning. Hegel warned us thus: "regardless of whether they have in and by themselves any *real meaning*". (WL I 276)

The least we ask scientists for is that they know what they are talking about, that is to say, that one understands them. That is the exigency of Hegel. Critics overlooked the fact that the verb *begreifen* means precisely 'to understand'. This is only one example among many. Ripalda translated it as 'to grab' and Lauer as *'to come to grips with'*. All at once they let escape the only exigency of the work of Hegel, the one that makes science a *scicentia*, knowledge and understanding of things. Strict logic consists in the exigency according to which we must go from one determination to another, *because* the first determination is not understood but the second one is. This second determination is at least closer to the true concept, which is the only thing that can be really understood.

I do not want to stop for long in this question which is merely interpretative. The key to this puzzle is that commentators suppose that one can understand something without *begreifen*. Even polemically, the following text rules out *ex professo* that possibility:

According to Kant:

> the concepts of reason must serve to *begreifen*, the ones of the intellect to understand (*verstehen*) perceptions. However, if the second ones are truly concepts, they are concepts then, by means of which one conceptualizes, and to understand (*verstehen*) the perceptions by means of concepts of the intellect would be a *begreifen* (WL II 407).

Everybody recognizes that the verb *verstehen* means to understand. This passage explicitly denies the existence of a *verstehen* that is not a *begreifen*, something that the critics who do not translate *begreifen* as 'understanding' assume.

I would like to quote one more text in regard to the subject of causality, for in those discussions the word influx is often used. Hegel holds that such word must be abandoned because it cannot be understood (as in *begreifen*). Should we translate *begreifen* in any other way, the argument would lack strength and even sense: "The relation of the influence is a relation of vulgar Philosophy. However, since one cannot understand (*begreifen*) how can material particles and qualities pass from one substance to another, such representation must be abandoned" (GP III 240).

Thus we could quote innumerable passages, but this is a question only of interpretation, and our real efforts concern demonstrations. For the rest, the first quoted text settles the interpretative question, since it explicitly denies what mistaken commentators and translators need to presuppose.

What we have said about the way out refutes the *coup d'état* apodictically. Such a *coup d'état* is ludicrous: it shies away from the consideration of the subject because, according to scientists, they refer to reality, to the object; but, as a matter of fact, they only refer to the words, without knowing what they mean and with no intention of finding the meaning out, for that pursuit will require to take the subject into account.

But by their pursuing that way out in their *particular* conclusions will make our argument even more incisive.

For instance, they say that science only knows the object and not the subject. But in that same phrase they are referring to something else than the object: science. They say they only know about the object, but they claim simultaneously to know about science, which is not the object. The key proposition turns out to be of the same nature as that of the liar's paradox: it is false because it is contradictory. Therefore, it is false that science only knows about the object.

They say that we only know the empirical, but therewith they make explicit a claim about something that is not sensible: knowledge. They build up a great thesis about knowledge and hold at the same time that they do not know anything about knowledge. They pretend to say nothing about knowledge but in fact they do: "the contradiction of a true which cannot be true at the same time" (WL II 441). They say that we cannot distinguish between scientific knowledge and unscientific knowledge, but they also say that we do not know anything about that knowledge.

They say science should only deal with what is perceived by means of sensation. But that sentence makes no sense if they do not define sensation. And sensation is definitely not perceived by sensation. Hence, they do not know what they are saying insofar they do not deal with something that is not perceived by sensation. Who knows what the phrase means with which they intend to carry out a *coup d'état* against the subject?

The most common thing to say is that we should deal with objects, not with the subject. But it is impossible to define 'object' without dealing with the subject.

In fact, they characterize every attempt to define object by means of the negation of the subject, that is to say, by means of contrast with the subject, and consequently they deal with the subject. And this is not something to be surprised about, since, naturally, the thing the subject knows more of is the subject himself; and, in every definition, he who defines should be more known than the *definiendum*. That is the crucial point of all the discussion, and scientists do not make up their minds to face it: the object (and the same happens with that of 'body') is one of the most abstract, devoid of content notions there are. All they can do is to define it in negative terms: everything that is not spirit, i.e. that which is not determined by itself, "the other of the spirit" (WL I 105).

It is obvious that they cannot define object (or body) as 'what is seen', for, to put an example, although electrons cannot be seen, and they are supposedly to be objects. Electromagnetic fields are not seen and one includes them among other objects. It would be pointless to say that object (or body) is what is seen in itself or in its effects, for one also sees the effects of the soul, in other words, the effects of the subject, i.e. a wave with the hand. The definition of the object would have to distinguish between object and subject, between what is corporeal and incorporeal in *itself* and not in its effects, for if it is only necessary that the effects are sensible, one could confidently say that everybody has a soul, and hence it would be false to say that science does not deal with the subject. But there is a stronger objection: if they say 'effects', they have to define that word and also causality; but causality, as Hume and Kant have demonstrated and as we will see in our next chapter, is not an empirical data and can only be defined in function of the subject. The fictional causality of the physics, which consists in the succession of one object after the other, cannot be applied here, for the so-called 'effect' is a phenomenon; the so-called cause, the electromagnetic field and the electron are not an object: they are not seen.

We have stepped into another variant of the coup d'état: we are dealing here with the material.

Let us repeat the objection: the material is a completely negative notion; its only content is that 'it is not spirit'; therefore, to designate a meaning to this new expression of the way-out, they would need to indicate what spirit is and deal with it.

It is true: to the coffee philosophy, lover of the banal, language indicates the opposite: the immaterial seems to be a mere negation of the material. And even theologians often incur in such a platitude. They forget what Hegel warns us repeatedly, "the language of ordinary life is made for the world of imagination" (WL II 357), not for the world of concepts which we are examining here. In regard to the world of concepts, Plotinus had already pointed out this very well: "The matter is only through abstraction of the other. That which remains when we take ideas aside, we say it is matter (Hegel GP II 459),

Hegel makes valid the same claim: "When one abstracts every determinations and every form of something, what remains is undetermined matter. Matter is simply an abstraction" (WL II 70).

It is useless that they say: we understand by matter something else. For they are not *understanding* anything when they say the word matters. So much so that they do not have the concept and they cannot define it.

It is an abstraction that has been laboriously fabricated by means of negations. For this reason Hegel says: "even physics have come to odds with matter recently" (EPW 389 A). If it is only a mere abstraction, abstractions only exist in the mind, and here we go again back to the subject, that which was tried to be deleted.

The process of rarification of matter in hands of scientists has continued after Hegel at a greater speed. For instance, Taylor and Wheeler explicitly recognize this:

"The best current thinking does not claim that particles are *not* built out of spacetime" (1966, 193).

The matter of scientists consists in space and time. Could there be a more abstract entity than this? Of the possible attributes of matter the only thing that has been retained is spatial location; the other ones have been relegated to the kingdom of shadows, because they would only consist in some relation with the cognitive organs; but what is it that fills space is not said; it remains as a mere negation; matter is what the spirit is not.

Historically speaking, it should be obvious that man would have never called something material if it was not by contrast with something that is not material. We simply would have not come to the idea of calling something material. The same happens with the pair 'exterior-interior': we would have never called the empirical world exterior due to anything else than the contrast with something interior. To affirm that only the material exists is one of those funny theses that in order to have meaning needs to be false. For we would not understand it if we did not know something, in contrast, by which the word 'material' makes sense. It is like with the thesis that states that everything is inexistent: we would not understand it if we did not know something existent.

While making a history of natural religions, Hegel correctly point out that in the ancient Orient "the interior and the exterior are not yet separated, nor the spiritual and the natural" (WG 269).

> To the prosaic point of view of nature —as that of the perceiving intellect—, the distinction between something purely natural and something purely spiritual, the intellectual observation of nature with which we look at the empirical is something that comes only very late afterwards, since it requires a superior return of reflection over itself. Only when the spirit has autonomously affirmed itself as independent of nature, only then nature presents itself to him as the other, as something exterior. (PR I 203s).

Humanity would have never called something material nor would have given this word any meaning at all, if the knowing subject did not know himself as spiritual. The negative morphology of the word 'immaterial' is a bait only taken by the inexperienced.

The attempt to define the material without the subject often states that the material is the extended, and the extended is what has parts. Despite that empirical impressions of taste and smell do not have parts and yet we call them both material —which makes more obvious that being material is not the same as to have parts— it is not clear that modern physics is willing to remain loyal to a definition according to which the material is that which has parts. If physics find some particle or phenomenon that does not have parts, they would still call it material without hesitation, which makes all the more clear that, to them and everybody that reflects a little upon the subject, the material means exclusively that which is not spirit and does not determine itself. Furthermore, they have already found that phenomenon or material reality that has no parts: what characterizes the quantum of Planck is precisely that it is not divisible. And no one doubts that it is material.

But the worse surprise for the authors of the aforementioned, traditionally-aimed attempt of definition is that the word 'part' cannot be defined but in function of the spirit (cfr. WL II 138-144; EPW 136 A).

In the material the whole-parts relation does not have *objective* meaning. Evidently, these two terms are correlative, the former falls in the definition of the latter and vice versa; nothing can be called a whole if it does not have parts, and nothing can be called a part without a whole. By no means two things can be called parts just for being juxtaposed; likewise, by no means something can be called a 'whole' just because we subjectively encompass two things by our eyesight as if they were a whole or parts of a whole. If each of them can exist on their own, they cannot be part of something else. What is objective are not 'as ifs'.

He who calls part a region, a blob or a stain that he imagines becomes a prey of his own imagination. If that region can keep itself in being, then it is an entity on its own account and not a part. That something is a whole does not consist in our encompassing it by our eyesight or our imagination, for we could encompass thus multitudes of different beings which do not have anything to do together and which *are* not parts.

There can be no real distinction between whole and parts, which is demonstrated by the above mentioned fact that the former falls in the definition of the latter and vice versa. If it was a real distinction, then the whole would not be a whole and the parts would not be parts. Being a part consists in being simultaneously identical and different form the whole, which is impossible in the material and can only be realized in the spirit. As we will see in our next chapter, the subject consists in the multiple determinations that he gives to himself, and does not consist in any other thing "the self is, thus, the completely simple and at the same time a multitude, a richness in itself" (EGP 277). In the realm of nature, only the organism can be —and, by the way, in a vulgar and diminished sense— fulfillment of the whole-parts relation; but the organic is not what is merely extended and material anymore.

It has been an illusion to define the extended as that which has parts, because the segments of the extended are nor parts in reality, each of them is an entity that exists on its own account; and our encompassing view cannot make material things be what they are not. Consequently, someone would be inclined to define the extended as 'what we perceive as extended', but aside from that evident circularity that defines nothing, the subject reappears in what we 'perceive', and that is precisely what they wanted to scare away.

Once the attempt to define the material as the extended is frustrated, one could want to define it as the sensible; but that would be to incur once again in the maneuver 'we only know the sensible', which has been deemed a fraudulent relative of the liar's paradox. In addition, as curious data, the sensible does not have a common sensible denominator or referent to the object; smells, sounds, and colors are among the most heterogeneous things we may find. The only common denominator between them is that they are perceived by the subject in a non-intellectual operation. Therefore, the general expression 'the sensible' does not have meaning without reference to the subject —that very same subject that they want to scare away. In sum, the material can only be defined as "the other of the spirit".

We have referred to the frustrated attempt to define it as that 'which occupies space'. It is obvious that such pseudo-definition does not tell us what the material is in itself, but only attends to an extrinsic relation that leaves us as ignorant as before. Since there is no intrinsic content, we would be rightly entitled to say that nothingness occupies space. The authors or this pseudo-definition have to provide a content if they want to distinguish between the material and nothingness. But there is something even worse: they would have to define space, an enterprise in which they fail because space is not an empirical data, and hence they cannot point out something with the finger as if that was a definition. If they define space as "what is occupied by matter", the circularity would be outrageous, and neither we nor they would know what the material is. Other aims to define space employ the word *extension*, but we already made clear that they cannot define it.

Leaving aside the variant of coup d'état that says 'we refer to the material', let us now move to the more interesting one: 'we refer to the real, we are not interested in the subject.'

The solution to this problem, which is, by the way, of utmost importance, lies in clarifying what does the expression 'the real' mean.

Otherwise, neither the authors of the *coup d'état* nor we would know what they are talking about. All metaphysics, considered as science of being, is here at stake. Scholastic Philosophers also believed that they could understand being without studying the subject, a supposition that makes their entire system perishable.

It would be a relapse, of course, to define the real as 'the empirical'. We refuted those who adopt that position. Electrons and electromagnetic fields are not empirical and yet the promoters of this *coup d'état* consider these organs to be real. Not to mention that there are empirical data in hallucinations and these very gentlemen consider that no reality corresponds to those data: it is clear that even they distinguish between reality and the empirical data. Consequently, it is false that they *understand* the sensible.

When they say that they refer to the real and that the subject does not interest them, it is obvious that they are uncritically assuming the recurrent and frivolous conception according to which the real can be defined independently from the self. This definition revolts furiously against itself as a tornado. They are actually defining the real in function of the self. The more independently the real is defined from the self, the more dependent the former becomes to the latter. Without the content of the 'self' it is impossible that something 'distinct from the self has meaning.' "For the self is that which explains and has a meaning by itself." (Ästh II 435).

There is a very similar definition, which deserves to be commented here because it detonates, with all its explosiveness, the problem we are dealing with. It states: the real is what is outside the mind, outside of thought.

The time has arrived to settle this question once and for all: since thought is neither something spatial nor a tub nor a salon, *the expression 'outside of thought' lacks any kind of meaning.*

Hegel says: "it is not outside of thought, but the thought from outside of thought".(GP III 145).

It is useless for them to say that what they mean is 'outside the head', for the encephalic mass and the hypothalamus are inside the head, and these gentlemen consider them to be real. Going back to the first definition, he who defines the real as what is distinguished from the self: does he suppose that the self is real or unreal? In the first case, it is false because the self is not distinguished from the self and yet it is real. In the second one, the definition thereby advanced tries to define the real in terms of the unreal, which is an absurd, for the *definiendum* would lack any characteristic content of the real, and we would still not know what "the real" means. It is obvious that the real would have to be defined in function of the real and not *vice versa*.

But against all their intentions, this recurrent conception hits the mark with the main point: it is impossible to define the real independently from the self. If the characteristic content of the real is contained in the *definiendum* due to the relation with the self, the only content possible is this: the self.

We could have conjectured this even *a priori*: there is nothing more real to the self than the self itself. Which other meaning could the terms 'real' or 'being' have other than 'self'? The first being we know is known to us by means of introspection: "Because in thinking the spirit knows itself as truly existent and real" (GP I 377).

This conclusion should not surprise us, for the present chapter has demonstrated that the cause of the concepts is not exterior and that their origin is not an empirical data. Moreover, the meaning of 'being' is not an empirical data. Neither color nor smell means 'being'. Hegel says: "one cannot see or smell the being". (GP I 517).

> […] the distinction between he who feels and what he has felt, between the touching subject and the touched object, as well as the relation which consists in that the object cause an *impression* in the subject, and this is *affected* by the object and the object is cause or stimulus, etcetera. None of these distinctions belongs to the point of view of sensation itself, but they all belong to a posterior reflection of the soul when it has determined itself as *self* and *spirit* – […] this a distinction that does not concern sensation as such (BS 542).

Our next chapter will turn back again to the subject of 'being', but here it is fit to say that Thomas Aquinas had already noticed that being is not an empirical data: "Although there is being in the sensible things, being as such, the formality of being, is not apprehended by the senses […], since they only apprehend the sensible accidents". (I Sent 19, 5, 1 ad 6um).

Kant also noticed this: "the being of a real object outside of me […] is never given by perception, but can only be added by thought to perception" (KRV A367).

Likewise, Aristotle said: "nor indeed can any of the 'intelligibles', e.g. Unity or Being, be an element; for these apply in every case, even to composite things". (Met XII 1070b 7).

And Plato as well stated: "Then knowledge is not in the sensations, but in the process of reasoning about them; for it is possible, apparently, to apprehend being and truth by reasoning, but not by sensation." (*Teethetus* 186 D). Also *Fedon* 65 C, and *Fedro* 247 C

All this authors only base themselves on intellectual honesty of analysis: should we strictly refer to empirical data only, we would have never come up with the idea of being, with the idea that something is real. Of course the object *is* real; but the senses do not provide us with such metaphysical and perceptive ideas.

It is inevitable to remember the warning we made against the stubborn illusion of those who believe that, for a concept to take form in the mind, it is enough that the object that lies before us *is* as that concept describes it.

There is no harm in saying that also the empiricists tell us that sensibility does not capture the being as such. For instance, Hume said: "...tho' every impression and idea we remember be considered as existent, the idea of existence is not deriv'd from any particular impression." (Treatise I, II, v i).

Believing the dogma that one can only demonstrate by means of empirical impressions, Popper says: "My thesis is that realism is neither demonstrable nor refutable." (1973, 38). The idea of Popper is the following one: if the real as such were a sensible data, we could demonstrate by sensible data that a reality corresponds to our subjective impressions.

Carnap is also explicit: "...the ascription of the property 'real' to any substance (be it matter, energy, electromagnetic field, or whatever) cannot be derived from any experience and hence would be metaphysical." (1969, 287)

In short, because the character of 'the real' is not an empirical data, the meaning of this term can only be obtained by introspection and identified with the existence itself of the subject. Real means: *like the subject*. What happens is that those, who believe that by a *coup d'état* they only refer to the real in order to do away with the subject, end up emitting meaningless sounds if they do not define real, and if they *do* define it, they flagrantly return to the subject.

In the light of this result, the authors who evade the subject will decide to take a strategic turn of one hundred eighty degrees and deny that the human mind is capable of knowing reality. This negation is void skepticism, the most modern shape that the Kantian thesis of the thing-in-itself has adopted. Their theses are 'there are no absolute truths' or 'everything is uncertain'.

We already said there is no reason to make a fuss about the second formulation: if science is quitted (which, by definition, is true knowledge), then the subject is avoidable. Besides, one should notice that the thesis 'everything is uncertain' is one of those funny propositions that, in order to have meaning and be understandable, needs to be false. In fact, the word 'uncertain' can only have meaning by contrast with something certain we know: therefore, automatically, not everything is uncertain. We would have never called something uncertain if it was not by contrast with what is certain. The thesis 'all is hypothetical' would also be uncertain, for hypothetical is something that is put to test in order to see if it is true or false, but a part of a test consists in a comparison with something that is not hypothetical, because, without something that is true, we would embark upon an indefinite process and the expression 'put to test' would lack any meaning.

The first formulation, 'there are no absolute truths', is more famous. In our first chapter we examined half-skepticism and we saw that scientific skepticism is the only radical one because it is the skepticism that thinks. However, the skepticism with which we are dealing now, although it believes itself to be very radical, is in fact an intellectual suicide and thus null and void. One must notice that, in real life, no one holds it, because every person is convinced of many absolute truths in the practice. But in theory this skepticism affirms with one hand that which the other denies, and hence it certainly does not say anything and becomes invalid. It affirms to know a truth about our cognitive capacity, and yet denies that truth is knowable. Formally speaking, the thesis 'there are no absolute truths' refutes itself: if it is an absolute truth, there are absolute truths and hence the content is false; if not, then its content may be false and it could be the case that there are absolute truths.

Such an opponent would try to elude us by saying: 'there is no absolute truth other than this one'. But he necessarily presupposes that I am capable of understanding what the word 'truth' means. Necessarily, he would be affirming that 'I am capable of distinguishing between my proposition and other propositions which I say are not absolute truths'. Hence there is another pair of absolute truths which are the conditions of possibility of the first one and which make it false. And let us not even discuss ulterior implications, for they practically encompass all the absolute truths of logic. For instance, if two propositions should be distinguished, they need to have a different subject, a different verb or a different predicate. Or even a more fundamental truth: there can be no proposition which is simultaneously true and false.

The only resort of this skepticism would be not to speak. But whatever is not expressed does not exist either to the theory or, in general, to the rest of men. It is void and null.

Hegel already told us: "So soon man speaks and there is a concept there" (GP I 336).

Notice that even one needs concepts; otherwise in order to deny something, one does not know what is denied. Even Carnap, who denies the cognoscibility of something real distinct from the empirical data, needs that this 'real' has a meaning; otherwise, all his famous theory about pseudo-problems is a confusing aggregate which lacks intelligibility.

8. Meditionism

By the way, the closest thing to not speaking is not to hurl words but numbers in front of a phenomenon. And meditionism consists in that: it states that anything that is not measurable does not exist to science. "Mathematical knowledge displays proofs of the present object as such, but not as such object was understood; it lacks the concept completely" (GP III 187).

Taking the meditionist thesis in a coarse and literal sense, it would be objectively ridiculous that a scientist, instead of telling us what he is supposed to —a task for which he would need concepts—, starts taking measurements left and right like a mathematician and babbling out numbers which do not make any sense. Anyone can start to count with all accuracy how many hairs does a cat have, how many times does the public sneeze at a concert, how many hats do people in the street have, and how many volutes does the smoke of a cigar make. But none of those things are science. It is clear that identifying science with measurements is, without further ado, mistaken. Not even if we add some systematization this would become science: I can count how many bald man there are in concerts and tennis games and how many hairs gray and brown cats have. I would be systematizing and taking measurements, but not making science.

We shall look more closely upon three things about the meditionist theses. First, why they say what they say; second, what is what they formally say; and third, what do they actually mean to say. The three things turn out to be untenable.

Why do they say what they can be explained by the fact that one ordinarily believes that the measurable is an empirical data. One simple (but important) reflection upon the most common thing to measure, namely, longitude, (which, by the way, falls into the definition of the other three basic measurements of Physics), suffices to show that such a belief is an illusion. As an empirical data, the longitude of this table depends of the distance and point of view from which one observes it. If I observe the table at 50 meters of distance the empirical data of the longitude would be completely different from the one I could obtain by an observation at twenty centimeters of distance. Since the distance and the angle can be infinitely varied, the number of empirical data which allegedly concretize the longitude of this table is infinite. Those are the only data which are completely empirical. The question immediately rises: to which of them do the meditionists refer? They answer: we refer to the longitude itself. My answer to this would: well said, but that is not an empirical data.

The reflection we just made is so obvious that, putting things the other way around, the measurement could seem to be more like a desperate attempt to avoid the empirical data, something very similar to the absolute space of Newton. Naturally, this attempt fails. Meditionists would tell us that 'objective' longitude can be inferred with all accuracy if we investigate the distance and angle at which we observe them, and thus we can do away with the innumerable optical appearances. But they forget that this distance is also a longitude, and that, in order to measure it 'objectively', one would need again to find out at which distance are we looking at it, and things would go on like this *in indefinitum.* It follows that the so-called objective longitude is not inferable.

And let us not forget the fact that the aimed objectivity would be to put everything in meters, but, what does 'meter' mean? If the meter of Paris, at once with all the other bodies, was ten thousand times bigger or smaller, we would not notice anything and there would be no difference. One may then ask himself: what does the aimed 'objectivity' consist in?

Both the empiricity and the objectivity of the measurable are a deceit; and that is the reason of existance of the meditionist thesis.

Let us examine now what it formally says: only what is measurable exists. In its excluding sense, this can be formulated the following way: 'there is no thing that is not measurable'. This thesis is the most typical example of propositions which are essentially indemonstrable and hence unscientific, for its corroboration would suppose to cover inch by inch the universe, across past, present and future, in order to prove that there is no single object which is not measurable. In its assertive sense, 'all what is measurable exists', is an affirmative universal proposition, which only needs a singular negative case in order to be proven wrong. Now, that case exists, and we have seen it: the surface. If there is one thing that is easily measurable that thing is a surface; one multiplies the base by the height and one ends up having the results in square meters. However, contemporary physics know well that surfaces do not exist; subatomic particles, reduced to wave packets, do not have surfaces. Consequently, meditionism is false both in its including and its assertive sense.

What the meditionists really mean is that they refer to the measurable based on an arbitrary decision. But well, even that is false. Evidently, it is not the same to them to have four units of time than four units of mass. What are measurable are the four units, and in that there is no difference. The difference is that in the former case we deal with time and in the latter with mass. That difference is not measurable but intelligible, and yet it is a difference of utmost importance to them. Therefore, it is false that they rely on what is measurable.

CHAPTER III

The Subject

We will expose in this chapter, three characterizations of the subject. The second will simply be a deeper analysis of the first one (which is the most decisive one), and the third will make explicit what was contained in the former two. In fact, the definition of the subject is one and only, but one cannot say everything with a single catch of breath.

Once we provide the two first characterizations, we will answer two apparent objections, one that is based on the concept of substance and another based on the concept of time, both which will be useful for us in order to weigh up the whole dimension of what we exposed. Likewise, with the same purpose, once we provide the third characterization, we will analyze three objections, one based on causality, another based on the ideas of natural and physical law, and another one still, based on the idea of necessity.

1. Concept

Here is the biggest discovery that man has ever made in his entire history: spirit is thought.

We do not say that the spirit *has* thoughts, as if the spirit consisted in one thing and had *accidentally* thoughts, or as if it were a non-thinking substratum, consisting in who-knows-what which was suddenly struck by thoughts. Neither have we meant to say that the spirit consists in its capacity or power of having thoughts. What is absolutely decisive is to realize that spirit consists in 'realizing' things, in being aware, and in thinking. Spirit is nothing aside from its own experiences.

Spirit consists in that ethereal and ideal thing which the idiots call 'mere ideas'. A couple whose love is not reduced to sex knows well that their reciprocal exchange of experiences, their communion, their marvelous mutual understanding and human flourishment are 'pure ideas'; but for this couple all of that is more real than the floor and the walls that surround them. That precisely is the spirit. Materialists are not mistaken when they say that the spirit is mere ideas; they are mistaken in saying that the matter to which they refer is more real than those ideas. The mere fact that they cannot define matter without the spirit demonstrates which of these two things is more real. In general, the objection against the thinness of the consistence of the spirit would only have strength if the objectors could indicate, independently from the spirit, what are they speaking about when they say matter. But we have seen that they cannot do this.

This unmatchable discovery was made by Aristotle: "being is to become aware or to think" (Eth. Nic. IX, iX, 9); "the mind is nothing before it thinks" (De anima 429b 32), "it has no actual existence before thinking" (ibid. 429a 24)

Also Descartes, independently from Aristotle, made this discovery. Hegel summarizes it thus: "his principle was: *cogito, ergo sum,* which cannot be understood as a syllogism in which the *ergo* denoted the consequence of the premises. It means rather: thinking and being are the same; it is a principle that is still valid today" (WG 915).

One must notice that, in fact, Descartes explicitly denies that his *ergo* indicates a deduction or an inference, for those operations would suppose premises. Hence, as Hegel correctly interprets it, Descartes can only be referring to the identity of thought and being.

But it was Hegel who demonstrated the truth of this discovery, and who gave it its authentic dimension in the formulations we are about to examine —the expression we have given it up to this moment, is still too poor—; Hegel was the first one to realize the importance of this for the sciences, because only on the grounds of this discovery it is possible to give meaning to words.

"The first thing one has to do with the concept is to stop believing that the concept is something we have, something we do within us [...] What we call soul is the concept; the concept as such becomes existing and it is spirit and consciousness" (PR I 220).

"What we call soul, what we call self, is the concept itself in its existence (Ästh I 175).

"The concept, insofar it reaches an existence which is free in itself, does not consist in any other thing that in the self itself and in pure self-consciousness. To be sure, I have concepts, determined concepts, but the self is the pure concept itself that has reached existence as concept" (WL II 220).

This is essential: if ideas and concepts are not to be identified with nothingness (something which would be utterly absurd), if some reality should be undoubtedly granted to them, then ideas and concepts must necessarily be identical with the reality itself of the spirit; otherwise, it would be impossible for the self to be aware of them. It would be impossible for the self to think if thinking itself (the ideas) was not identified with the self. If, in the operation of understanding, the self was not identified with the concept, the former would need within itself another concept to understand the former concept, but then the first concept would be superfluous. "Self-consciousness is the category that becomes itself conscious" (PG 284).

> If we remained in the mere imaginary representation of the self —as imagined by our ordinary conceptions—, it is a simple thing which is also called a *soul*, to which the concept is *added* as if it were a possession or a property. This imaginative representation that refuses to understand the self and the concept cannot be useful in order to make easy or possible the intellection of the concept (WL II 222s).

"In rational psychology, which is an abstract metaphysics, the soul is considered not as spirit but as a merely immediate being, as *a soulthing*." (WL I 220)

That thing which spiritualists called the soul, insofar it did not consist in the activity itself of thinking and understanding, was not spirit: it was a bizarre species of matter they had the whim to call immaterial. They could not define it, they did not have the concept of it, and they only *imagined* it. Accordingly, it was extended, for only the extended is imaginable and hence was not spirit. For that reason Hegel says that they refuse to *understand* and remain only in the imagination. The moment one says that he understands (and is not imagining things) is the moment he knows that the spirit consists precisely in that.

"The concept of self, as the act by which the thinking itself becomes an object for itself, and the object as the one self, are absolutely the same; without that act the self is nothing." (GP III 427)

A substance without act or before the act, which consisted in something different than the act itself of understanding, would be what is vulgarly called matter and would not be spirit.

"The spirit is that; not being immediately, but only as an object to itself" (PR III 14).

> Definitively, the spirit is not immediate; it does not exist in the way of immediacy. Natural things are immediate; they remain in the immediate being. However, the spirit only exists in so far it suppresses its immediate being. If it only is, then it is not spirit; for its being consists precisely in being mediated to itself as a spirit which is for itself. The stone is immediate, completed. (PR I 70).

"I exist as a spirit insofar I have knowledge of me" (GP I 51).

"The spirit is not natural; it is only that which it knows how to do" (GP II 494).

"The spirit itself is only this perceiving of itself" (GP III 193).

"The spirit is nothing else but a producing that becomes an object for itself" (GP III 427).

"The spirit is having oneself as an object" (PR I 65). "If we do without thought, the soul does not longer exist" (GP II 48).

"*Self* means simply *to think*. If I say: *I think,* this is a tautology. The self is perfectly simple. *Thinking* is the way I *am,* and this is *always* so (NH 164)."

"The child is only spirit in itself; he is not a fulfilled spirit; he is not real as spirit; it only has capacity, the potentiality of being spirit, of becoming real as spirit" (PR III 204).

"Although the embryo is *in itself* human, he is not a man *to himself*; to himself he would only be once he has a formed reason that *makes* itself what *it itself* is; only then there is man" (PG 22).

Let us draw our attention to that which Hegel notices:

"If this has seem *new* in recent times; that has its cause in the ignorance of the concept of Aristotle" (GP II 158).

It is one of the most impressive and sad facts of the history of thought that rational psychology (which is an abstract metaphysics), despite of proclaiming itself as a discipline which followed Aristotle, had let escape, in fact, the most important discovery made by the Greek philosopher. This false Aristotelism accepted (without understanding it) the notion of pure act in its application to the divine spirit, but not in regard to the human spirit. The discussion is not about a particular application, but rather about knowing what the spirit is in itself: "there are neither two kinds of reason nor two kinds of spirit" (PR I 43).

Both the human spirit and the divine spirit are spirit. What we want to know is what does being a spirit consists of. In what was a clear misunderstanding of the problem, scholastic and rationalist philosophers believed they could define the generic notion by means of a negative route: the immaterial. But we saw that it is impossible to define the material, except as the "other than the spirit", that is to say, as negation of the spiritual in terms of the spiritual. What is first understood is the spirit; if not, nothing is understood: "For the subject is that which has meaning by itself and which explains itself "(Ästh I 435).

In order to distinguish between the divine and the human spirit, scholastic philosophers and the followers of Wolff would first need to define spirit as such and for that purpose they turned to matter, but in order to define matter one needs to define spirit as such. The circularity is manifest, and there is no way out of this predicament. In one word, they did defined nothing.

"If we think the spirit as immediate, simple, quiet, it is not spirit; the spirit is essentially this: being in activity. Furthermore: the spirit is the activity of manifesting itself" (PR I 65).

"To such a degree this is its substance itself that we cannot speak of it as an invariable subject that makes or operates this or that, as if the activity was fortuitous and some kind of situation outside of which the spirit had consistency; its activity is rather its substantiality, the activity is its being" (BS 528).

"That movement itself is the self" (PG 22).

Scholastic philosophers did not understand that act means acting, moving, making, operating, effectuating. They thought that act meant feature, characteristic, quality, determination (a 'perfection', so they called it). But the spirit is the act itself of manifesting itself, the act of making an object of itself: "the spirit itself is nothing else but this perceiving itself" (GP I 93).

It is not that the spirit exists first and then becomes aware of itself. In what could the existence of the spirit lie if it were not in taking consciousness of itself? This 'awareness' constitutes precisely the species of reality that is the being of the spirit, the peculiar kind of reality called spirit. If we have used in the previous chapter the term introspection in its vulgar sense, here we will not employ metaphors anymore. What we are dealing with is the pure act called consciousness.

2. Self-determination

Another way to express what we have been saying is this: the spirit determines itself. If the very same act of understanding and knowing are equal in it, and that act is carried out by itself, then the spirit gives itself the being. We already said with Hegel: "The spirit is not natural; it is only what it has made of itself" (GP II 494).

If spirit is not determined by itself, it is not spirit but other thing; for the spirit is not immediate, it does not exists in the mode of being of immediacy; it essentially consists in being mediated by itself.

It is not that something first exists which will later be perceived. That something would be the self, for no one can hold that the spirit exists without a self; but the self consists precisely in perceiving itself. It would be absurd to say that the self exists before the act of perceiving. "The spirit exists only as its own result" (VG 58), since the spirit "is only this perceiving of itself" (GP I 93).

"To produce itself, to make an object to itself, to know itself —those are the things the spirit cares about. The natural things are not for themselves; that is why they are not free" (VG 55).

When Marx, as a materialist, greeted joyfully the Hegelian theory according to which man makes himself, he did not notice that Hegel posed the pure essence of the spirit in contrast to everything material, something, and he does this exactly in the sense in which the material cannot make itself: self-determination (or, if one prefers, free will). We will get back to this point in a different context, but for now it is obvious that in order that self-consciousness exists there must be an interpellation from another self which has been self-conscious since always. "The Greeks did not have the idea according to which man was made to the image of God" (WG 577). "That the spirit is that which it makes itself being constitutes only one side of the subject" (WG 577).

This is what now occupies us. If the being of the spirit is the same act of understanding, the spirit makes its being, and hence the best definition of spirit is self-determination. In our previous chapter we saw that the only possible definition of 'object' and 'body' is what is not determined by itself. "The *only* determination of the spirit, in which *all the other ones* are contained in its *freedom*" (NH 58).

> [...] freedom is the substance of the spirit. It is absolutely clear to everyone that the spirit, aside from other properties, posses freedom as well; but philosophy teaches us that every property of the spirit has consistency only through freedom; every one of them is only a means to freedom; every one of them looks for it and produces it. This is the knowledge of speculative philosophy: that freedom is the only true thing of spirit (VG 55).

> The highest determination that thought can find is *free will*. Every other principles such as happiness or the welfare of the State are undetermined to a greater or a lesser extent; on the other hand, free will is determined by itself, for it is no other thing that determining itself (WG 920).

What these three texts express is something extraordinarily important which is in itself obvious: there is nothing more intelligible than self-determination. It follows from this that all the contents of the different concepts derive their intelligibility from self-determination. In other words, only in function of self-determination, which is the essence of the spirit, it is possible to give meaning to the concepts.

That self-determination is the best definition of spirit was something that Plato already said: "What is the definition of that being that we call soul? Do we have any better one than the one we just mentioned: the movement capable of moving itself? (Laws, X 896A). Evidently, Plato was also aware that such content is also the most intelligible one.

It is noteworthy that for Kant, in spite of all his skepticisms, this content is also the most intelligible one:

> Since its reality has been demonstrated by an apodictic command of practical reason, the concept of freedom constitutes the touchstone of all the edifice of a system of pure and even speculative reason; and all the other concepts (namely, God and immortality), which as mere ideas do not have any secure grounds within speculative reason, are added to the concept of freedom and by means of it obtain consistency and objective reality, that is to say, the possibility of them is proved by means of the fact that freedom is real (KPV 4s).

To the two characterizations of spirit we have given —namely, that spirit is concept and self-determination— we will add later a third one: interpersonality. But first we need to take a look at two extremely important concepts from which, apparently, two objections rise against what we have said: the concept of substance and the concept of time. The purpose of examining them will not only be to dispel objections; on the contrary, we will examine them in order to see that self-determination not only is more intelligible than those concepts, but also that it constitutes the only way in which we can give meaning to both.

3. Being

First, we must see this in regard to the concept of being. There cannot be any better preparation for our analysis of the concept of substance. Traditionally, one affirms that the substance is that what truly *is*.

To believe that 'everybody understands' what the word *reality* means is an illusion that was dispelled in our second chapter, insofar 'understands' means to consciously have the concept. Perhaps the cause of this illusion was that some thought that being is an empirical data, and hence that it was enough to point with the finger at some empirical fact to give meaning to 'being' and 'reality'. But we have demonstrated that the senses do not seize being as such, and that it is not possible to define reality as the empirical.

Nevertheless, the other cause of the aforementioned illusion has been, without any question, in order to figure it, it is enough to define being as 'that which is distinct from nothing'. This is a very similar mistake to that in which the Scholastic philosophers incurred by defining the spirit as 'what is distinct from the material'. The most elementary reflection makes us see that, if we define being in function of nothingness, the *definiendum* does not have any content proper of being but only the characteristic content of nothingness, and consequently, against our intentions, we would be conceiving and identifying being with nothingness. Despite how much we add negation ('distinct from…') to nothingness, all the content is negative. We are not seizing the being, for if there is something positive, that thing is the being.

That is precisely what happened to Parmenides. And all the scientists and almost all the philosophers that followed his path.

Evidently, there is nothing more immutable than nothingness. This is the reason why Parmenides said that the being is immutable.

Evidently, nothingness does not increase or decrease. That is the reason why Parmenides said that being does not increase or decrease.

Out of this apocalyptic confusion between being and nothingness —which Hegel mocks at the beginning of the *Science of Logic* explicitly referring to Parmenides in the first note— sprung the physical law of the conservation of matter, a law according to which the quantity of matter in the universe does not increase or decrease. It is obvious that the physics are incapable of defining matter; they do not *understand* anything when they use that word; but when there is no content in the mind and there is a total lack of determination, the content is null. That is exactly what happened to Parmenides and all who thought that being can be defined in function of nothingness.

About that apparent being Hegel rightly says that "nobody can tell what it is" (WL II 241).

They justify themselves by arguing that it is a simple and primitive notion; but such a statement does not add any content to it; we remain thereby in nothingness.

"According to those so-called authors, unity, reality and alike determinations are *simple* concepts, only because logicians could not discover their content and were satisfied only by having a *clear* concept of them, that is to say, no concept at all" (WL II 255).

Although they use other words as the grammatical subjects of their propositions, all what the principle of Parmenides and the principle of conservation are saying is a tautology: nothingness is immutable; nothingness does not decrease or increase. Since they are not able to define the verb which they place as a grammatical subject, since the only thing they speak about is lack of content, it follows that the subject is nothingness.

In our understanding, the impossibility of defining being in function of nothingness should be obvious. It should be manifest that everything that has content cannot be defined in function of what lacks all content. Our second chapter showed that the meaning of being cannot be an empirical data. Thus the origin of that concept could not have been sensation. Therefore, this conduct could have only been obtained by self-consciousness. But by self-consciousness what we discover is self-determination or, following the marvelous Platonic formulation, "movement (that) is capable of moving itself". Far from being immutable, being is movement; continuous novelty, continuous production of new determinations. In what could the act of existing consist if not in a movement of self-fulfillment and of giving oneself determinations?

When we predicate being of other things, we do this in a diminished and derived way. The true being is to determine oneself. As we have seen, things other than this can only be defined as "the other of the spirit". That is the reason why Hegel says that the truth of being and nothingness is *werden*, to become, to turn into, the movement, the process and the action of determining oneself and of acquiring determinations. The only intelligible content lies therein: this is what Hegel means when he says that a certain concept is the truth of another concept.

Neither Parmenides nor the principle of conservation of the physics speaks of reality, but only of an abstraction of which they have deprived all content. The content of the word reality is change itself, to create incessantly new determinations; whoever does not refer to that content cannot use the word reality, for not even he knows what he is talking about.

4. Substance

Let us examine now the substantialist objection that, apparently, can be raised against the two characterizations of spirit we have made.

To be sure, we will not examine the irreflective objection of those who, having accepted that God is pure act, figure that the immortality of the human spirit could be jeopardized if the spirit itself did not consist of a permanent substratum which is independent from its multiple acts of thought and will. If being pure act as God does not entail that danger, the same goes in the case of man.

This 'something else' that the substantialist theory desires is, without question, the self. They think it is obvious that the self remains identical through all the changing determinations. But, evidently, a mistake lies in here, because the self consists precisely in the act of perceiving oneself; it is even tautological to say that the self does not exist independently from the act of self-perception. The self is something purely ideal, whose reality consists in being perceived. It is here where imagination plays tricks with the objectors' minds: they imagine that the self is some kind of material rock that remains unaltered through several acts of will and thought. But the self is something purely ideal, a mental content that we call self-consciousness. The objectors believe that, in the act of self-consciousness, the rock called the 'self' is perceived but it doesn't perceive; this is why I say that they imagine a material self, because this self cannot know: it can only be known. They do not seem to understand that self-consciousness implies, by definition, a perfect identity between the object known and the knower: it implies the absolute purity in which knowing is the same as being known.

It would be useless, of course, that they called *immaterial* that 'something else' which they posit. We saw that it would be circular and hence null and void to define the material in terms of the immaterial, that is to say, to define the immaterial as the negation of the material. In order to define the immaterial, it is necessary to refer to the spirit and to deny the other. Consequently, the spirit has to be understood directly. But we also saw that the only possible meaning of spirit is the acts of will and reason. Therefore, if this 'something else' is immaterial, it has to consist in the acts of willing and thinking, and hence cannot be something different from will and reason.

If they contend that we do not understand the difference between *ens quod* and *ens quo*, I respond to them that neither do they, because the thing they refer to is only Latin gibberish. Leaving aside verbal curtains of smoke, what matters is to know if the spirit really distinguishes itself from its acts. If it can be distinguished, nobody can point out what does it consist of without reducing it to something material, despite *they call* it immaterial.

But let us give a closer look to the epistemological root of the problem. Fortunately enough, half of the journey is already past behind us, for those who despise the concept of substance and those who have a more positive image of it accept that the meaning of the concept is not empirical. All we perceive with our senses are accidents of the substance, not the substance itself. Hence, the origin of this concept is not sensation but self-consciousness or, if one prefers to say this otherwise, introspection. Now, all introspection is an *act* of introspection; all self-consciousness is an *act* of self-consciousness. Our objectors cannot refer to a concept of substance without act, because they have never perceived a substance without act, and only in such a perception could that subject have had its origin. The concept of a substance that is radically different from its accidents does not exist. They do not have the *concept* of a substance which does not consist in some acting. By the mere fact of perceiving the substance they have perceived it in act.

Such an act is the true substance. They have an *image* —a very blurry one, indeed—, because, if possible, they would strip it off from all determinations, since every determination would be an accident. And who knows? Perhaps they imagine a white mass in order to take all the color away, or perhaps they imagine it without form, since all form would be an accident. But, evidently, it must be extended; or else, it would not be imaginable. And from this follows that what they imagine is not spirit! In addition, the substance would not differ from the accident, since extension is an accident, and, let us not forget, whiteness is an accident too.

We have just expressed what is decisive: the act itself of the spirit is the true substance.

Substance is commonly defined as that which exists in itself and not in something else. Now, one cannot understand how something can exist in itself if such thing does not give to itself the determinations of its existence —and that is self-determination, the definition of spirit, "for the essence of freedom consists in being by means of itself what it is" (Ästh I 468). If its true realization is not spirit, the word *substance* does not have any meaning at all.

In considering that the spirit and the concept are exactly the same thing, Hegel expresses thus a decisive point: "The concept is the truth of the substance" (WL II 214). We pointed out before that when Hegel says that a notion is the truth of another notion, what he is really trying to say is that the latter only becomes intelligible in the former. By the way, this is the key to *The Science of Logic*: all the determinations that are studied in the first two books —among which, we find the concept of substance—, have their truth in the spirit or the concept, which is the object of study of the third book. Only if the realization of those determinations is the spirit itself, they become intelligible; otherwise, they cannot be understood.

"Subjectivity is even the absolute form and the existing reality of substance." (Rph 152); "this infinite reflection in itself [...] is *the plenitude of the substance*. But this plenitude is not the substance itself, but rather something superior: *the concept*, the being." (WL II 216).

As such, *substance* and *accident* are correlative terms. It is impossible to understand one without the other in the definition, for the accident is commonly conceived as that which finds its existence in a substance. In addition, the etymology of substance entails essentially something that is underneath the accidents. It is a very vulgar and coarse spatial conception that which speaks in terms of 'upside' and 'down'. It is evident that one would have to reject it, but even assuming it we could argue that, if the notions of substance and accident imply one another, the word *accident* turns out to be very deceptive because it denotes something that is accessorial, but we know that the substance cannot exist without the accidents and hence the accidents are not accessorial. "The substance is the totality of the accidents" (EPW 151), "the substance is the necessity of its accidents" (NH 21). Now, this only comes about fully in the spirit, for the spirit is a substance that consists in its very acts of will and thought. The spirit is precisely a substance whose accidents are not accidental.

"This movement of accidentality is the actuality of the substance as *a peaceful producing itself* [...] and the accidentality is the substance itself (WL II 186)."

"Only the whole has reality as such" (PG 476); therefore, substance and accident, imagined as distinct beings, turn out to be two abstractions carried out by the abstract intellect in its pursuit of immobilizing reality in order to supposedly understand it better. "To distinguish between the simple identity of being and the movement of the accidents in the substance is a way of sheer appearances. The first thing is the uninformed substance of imagination [...] which has no truth at all" (WL II 186s).

5. On the method

We cannot postpone any longer a fundamental reflection upon the method which we have followed, for the way by which we have given meaning to 'being' and 'substance' exemplifies, paradigmatically, the way all concepts receive their meaning. Our second chapter showed that it is impossible to grant an empirical origin to the meanings in the mind. On the contrary, a search for those meanings always leads back to the knowing subject and the spirit. Hegel says: "We cannot be satisfied with the empirical way; we should rather posit the ulterior question, how does the spirit arrive to the content, that is to say, the spirit as such, we or the individuals or the peoples?" (VG 53). It is not enough to know that the meanings have their origin in the knowing subject himself. It is still necessary to clarify how the subject shapes and constitutes each of them.

In order to answer that question, Hegel turns to Kant's enduring contribution of the first *Critique*: without the 'I think' there cannot be consciousness of the object and without self-consciousness there cannot be consciousness.

Hegel says:

> One of the profoundest and most certain intuitions that are found in the *Critique of Pure Reason* is to discover that the unity in which the essence itself of the concept consists is the original synthetic unity of apperception; that is the unity of the 'I think', the unity of self-consciousness. [...] Object, Kant says there, is that in which the concept of the *multiple* from the given sensible intuition *is unified*" Therefore, this unity of consciousness is the only one that refers the representations to an object, and hence makes them have the validity of an object, something on which even the possibility of the intellect depends (WL II 221).

In other words: phenomenologically speaking, there cannot be any object without subject; it is impossible that the multiple, intrinsically chaotic and incoherent empirical impressions are unified in the making up of an object if there is no self or an 'I think' before whom and for whom the impressions make up an object.

An object is that which is independent from the self. But notice the following: in order that the expression 'independent from the self' has meaning, it is necessary that the term 'self' has already a meaning.

Hegel also expresses this idea in his own terms:

> The truth of consciousness is self-consciousness, and the second is the explanation of the first, so that in the facts each consciousness or perceiving of other object is self-consciousness: I know about the object insofar it is my object (it is a representation of mine), therefore, I know about myself in that (EPW 424).

Kant's ground-breaking discovery is true and has an unmatchable value but it is still incomplete. According to Kant, it is necessary that the 'I think' accompanies our awareness of the object; it is necessary that self-consciousness accompanies consciousness; otherwise, the latter could not exist. But Hegel makes this sharp observation: "The I must 'accompany' it; what a barbaric expression" (GP III 343).

Hegel was right. What can that companionship or juxtaposition could possibly mean? A profound analysis shows that consciousness-of-the-object and self-consciousness do not simply go hand in hand. Undoubtedly, Kant was right in saying that, without self-consciousness, the concept by which we gain awareness of the object cannot exist, but both things constitute an inseparable unity. They do not accompany one another; but rather, they make up one and the same experience. "I know about my object and I know about myself; these two things are not separable" (VG 54).

If they are not separable, something logically follows from this: "The thoughts and representations that I make mine are given the content of that which I am myself" (PR I 22).

The content of the concepts by which I gain awareness of the objects has to be the same content of self-consciousness. This is exactly what happened to us with the concepts of being and substance. And the same must go for all the rest of the concepts.

If the meaning of the concepts cannot be an empirical data, the meaning must lie in the subject. But if the awareness of the object is inseparable from self-consciousness, the content of the former has to be the same as that of the latter. "The truth of consciousness is self-consciousness" (EPW 424). We could have demonstrated this at the beginning of this chapter by recurring to Kant's indisputable contribution, but without some examples it would have been quite difficult to understand. "The knower [...] hast the concept of the essentiality of the objective world entirely" (WL II 438); "we are, as a self, the base of all our determinations. Insofar the object is thought, it receives the form of thinking and transforms itself into a *thought* object. It is equaled self, that is to say, it is thought" (NH 164).

If it is not equaled to the self, it cannot be *understood.* That the true of consciousness is self-consciousness means that only the second is understood.

"Reason is the supreme unification of consciousness and self-consciousness, that is to say, of the knowledge of an object and the knowledge of itself. It is the certainty that its determinations are objective determinations of the essence of things as they are in our thoughts. It is, in one and the same thinking, with all the power of the certainty of itself, subjectivity, as being or objectivity" (NH 122).

We will get back to this later. However, it is important to say here that the abstract intellect —in contrast to reason— considers these two things separately, and hence it cannot understand.

"Abstractions of the intellect are not ideas; the idea contains essentially the point of unity of subjectivity" (WG 588).

"The objectivity hast the object in the concept, and this is the unity of self-consciousness, to which the object is incorporated; therefore, its objectivity, that is to say, the concept, is nothing else than the nature of self-consciousness, it has no more elements or determinations than the self itself" (WL II 222).

In that sense "the intelligence is *acknowledging*" (EPW 465), and if the intelligence does not recognize itself in the object, it would not understand anything. Because of this, Hegel says that the reason of scientists and rationalists —which believes that its only activity consists in observing the phenomena— deludes itself completely about what it is and does.

> As observing consciousness reasons addresses things, truly believing as if they were empirical things opposed to the self, but what it really does contradicts such a belief, for it *knows* things, it transforms sensibility of them in *concepts*, that is to say, in a being which is simultaneously a self (PG 184).

"To know things means precisely to eliminate what is alien and bizarre in consciousness, and thus it is a return of subjectivity towards itself" (WG 730).

The content of all concepts must consist in '"vital and spiritual situations" (WL I 335), for that is the content of self-consciousness and the self. We have just demonstrated this in regard of the concept of being: in what could the act of existing consist of if not in the act and movement of self-realization, in giving oneself determinations and in creating from nothing new experiences? According to Tomas Aquinas, the self is act. And the same goes for the substance: one cannot understand in what sense something can exist in itself other than in giving determinations to itself. Analogously, the meanings of force, cause, infinite, identity, and distinction are also spiritual and vital. One could contend this absolutely central thesis of Hegel only if one is able to provide an empirical meaning to those concepts, but we will see that such thing is impossible.

In order to complete our study of the substance, let us quickly draw our attention to the fact that those who deny the existence of the soul are being deluded by a concept of substance that lacks all kind of meaning: we made clear already that such 'something else' imagined by the substantialists objectors would be matter and not spirit. A good example of the negation of the spirit based on the void concept of substance is provided to us by anthropologist Alfred Kroeber:

> Dr. Häberlin puts in the foreground of Wundt's thought the idea that mental phenomena possess, through their immediacy to ourselves, an actuality as great as that of physical phenomena. This actuality must be consistently distinguished from substantiality. Only on the basis of this distinction is it possible to speak of the existence of a 'soul'; but in the sense of an actuality there is no denying this existence. (1952, 53).

In order to distinguish between the acts, the mental phenomena and the *substance*, Kroeber would have to give another meaning to the word *substance*. But we have seen that such content is an image of the fantasy and not a concept, because there cannot be a concept of a substance which is different from its own acts. When Kroeber grants the existence of actuality he has granted everything. By the way, this is precisely the true and imperishable substance, but we will get later to this point.

6. Time

In appearance, the following objection raised by our imaginative objectors against our characterization of the spirit, as the act of thinking and self determination, is conclusive. The objection can be formulated as a question: during the time that the spirit does not think or is conscious, does the spirit cease to exist?

The answer can also be formulated as a question: what makes them think that such time exists?

It is not the case that the soul does not exist during the time it does not act, but rather that such time does not even exist.

Time is not an empirical data. It can only be known by self-consciousness, but self-consciousness is an act; therefore, it is impossible to perceive time without action. Our objectors presuppose time illegitimately with no grounds whatsoever.

Aristotle made these conclusive remarks: "time is something related to movement" (Phys IV 219a 9); "it is evident that there is no time without movement and change" (ibid. 218b 34).

If there is neither movement nor change in the soul, there cannot be any time in it. Our objectors speak of an exterior time which they believe to find in the movement of the stars and the earth. However, such attribution is completely mistaken: it lies entirely in the Newtonian myth of an absolute time that comprises the universe. Apart from being a myth, the absolute time has been refuted by the science of our century.

Such attribution, as I said, is ungrounded, since we are not entitled to attribute to the soul a time of which it has no experience. The soul could not care less of what happens to the celestial bodies and other things alike.

The absolute time has been refuted. If Einstein demonstrated something that thing was the relativity of simultaneity. Although not many physics think about this, such relativity means that there is no cosmic 'now' that penetrates the entire universe. This inexistent cosmic 'now' is the absolute time that our objectors want to introduce into the soul from the outside, i.e., from the exterior things. As we said, however, the soul has its own time, a time that, as any other time, is essentially 'something about movement', because 'it is evident that without movement and change there is no time'. Therefore, there can be no time in the soul without experience. This time had already been discussed in the study of the 'time' that elapses (?) between the death and the corporeal resurrection, but the most lucid theologians suspected wisely that this was a pseudo-problem, for such time is extrinsic to the soul and has nothing to do with it.

It is important to mention here —as Hegel already stated it (cfr. II 1)— that the absolute time is nothing more than an abstraction. By means of experience we can only know concrete time and real events. Just like absolute space, so absolute time is a mental construct completely separated from experience. Such an abstraction has been constructed by men in order to determine a universal, but this they do by eliminating the concrete times which each of us experiments and which are the only ones real. This has been stated by the sociologist Niklas Luhmann:

> Time is gradually being abstracted, separating itself from the empirical world, it loses its intrinsic dependence from the habitual succession of events and things, it breaks the knot which bounded it to the vital rhythms [...] It transforms into an abstract *continuum* of points of time in which everything can move itself according to 'laws' and 'systems' which are not time (1976, 56s).

We will demonstrate afterward that this abstraction in fact is not time. Before we do that, however, it is important to emphasize that

Newton himself, in his *Mathematical Principles of Natural Philosophy*, was not unaware of the fact that he was constructing a big abstraction that was different from what he calls 'vulgar time', which is the one that we truly experiment and that actually exists. Newton says there:

> Absolute, true, and mathematical time, of itself, and from its own nature, flows equably without relation to anything external, and by another name is called duration: relative, apparent, and common time, is some sensible and external (whether accurate or unequable) measure of duration by the means of motion, which is commonly used instead of true time; such as an hour, a day, a month, a year (1977, 8).

As we will later say in a more detailed analysis, time cannot be an empirical data, not even relative time. But what is really amazing of this ghostly construct is that Newton does not even try to define his absolute space or time —something that he could not do because those things were only the products of his own imagination. On the one hand, he justifies his not providing a definition of space and time by saying that everybody knows what they are; on the other hand, he insists in distancing himself from what everybody understands by space and time, because such vulgar ideas do not seem reliable to him:

> I do not define time, space, place, and motion, as being well known to all. Only I must observe, that the common people conceive those quantities under no other notions but from the relation they bear to sensible objects. And thence arise certain prejudices, for the removing of which it will be convenient to distinguish them into absolute and relative, true and apparent, mathematical and common (*Ibid*).

What a resort! First he says that he does not need to define space or time because the laymen know what they are, and then he rejects the conception that the laymen have of them. It is evident that Newton needed some way out: he was unable to define absolute time for the simple reason that it does not exist.

With the above we have refuted the objection raised against us, since it rested entirely on the myth of absolute time. But we need to clarify, once and for all, the concept of time: the central thesis of this chapter still has to be confirmed and studied in more depth.

It is obvious that our objectors —as innocent as only they can be— *suppose* that time is continuous. Newton would also like to capture such continuity when he says that time 'flows' and that it is 'mathematical'. Now, in the sense that Newton and our objectors conceive it, continuity is not a data of our psychological or empirical experience. It could not be the former case, because no one can be incessantly having acts of self-consciousness, but it could not be the latter either, because even if we gratuitously suppose that physical time is continuous, our perception of time is not uninterrupted: no one can be carrying out acts of perception incessantly.

In our century something both curious and incongruent has happened: even though physics have rejected absolute time for not being an empirical data, they still believe in a relative time that possesses the same dubious continuity which is not and cannot be an empirical data. And they do exactly the same thing in regard of absolute and relative space.

The highly dubious continuity we have alluded to is the mathematical one. Following Bergson, Capek has very well remarked that the mathematical is a "disguised discontinuity" (1973, 321), for it supposes the existence of several different units —called 'points' in the case of space and time—; one wants them to yield the impression of continuity because one thinks of them as standing in a juxtaposition, but in fact they are external from one another. That is called *contiguity,* not *continuity.*

As Poincaré noticed, one demands a strong nexus between the elements of the *continuum* for that is what really makes it a whole; in the real continuity, the point has no priority over the line, but rather the line has priority over the point. According to its true concept, the *continuum* is unity in multiplicity; but in mathematical continuity "only the multiplicity subsists; the unity has disappeared." (1909, 30)

Hegel had observed this long time ago: "To the *imagination* that lacks the concept, continuity easily transforms into *juxtaposition,* which is an extrinsic relation of the units between themselves, in which each unity preserves its absolute discontinuity and exclusion" (WL I 181).

The inescapable conclusion —a conclusion which requires an intellectual honesty of which physics commonly lack— is that the time and the space of the physics are contradictory entities:

> It is not contradictory in imagination that one puts points in space, moments in the continuous time or the now of time within continuity length (day, year), but its concept is contradictory. The identity within itself means

> absolute continuation, extinction of all distinction, of the negative, of all the being-on its-own; on the other hand, the point is the pure being-on-its-own, the absolute distinguishing itself and suppression of all identity and coherence with others. Now, both things are affirmed as one in space and time; therefore, space and time are contradictions (GP I 307).

"The space is within itself the contradiction of the indifferent exteriority and the continuity without distinction" (EPW 260).

Notice well that this is not only about absolute time and space. We are not rejecting certain entities on the grounds that they cannot be made objects of our experience. This is much more about all sorts of space and time in Physics. What Hegel demonstrated was that space and time in Physics are contradictory in their very same concept. In the name of this strictly contradictory time, the objection was raised against the spirit as act of self-consciousness and self-determination.

Since mathematical continuity is not true continuity, it is now important to emphasize *that the time of the physics is not time.*

Jerome Rothstein is very brave to admit that the empirical data do not suffice to build up time. He explicitly states the following: "It is hard to give any sense at all to the term 'past' without presupposing memory or a record of some kind" (Bastin, 1971, 293). Resorting to memory is to witness the reflection of the subject upon himself, that is to say, to witness self-consciousness; so we are in the true time, identified with our self. In contrast, the alluded time of the physical world is simply not time. I don't know why it has taken so long for an exceptional physic like Rothstein to realize that, since Hegel had already demonstrated this with absolute clarity. In regard to the three dimensions of time, Hegel explicitly made this remark. "In addition, one cannot distinguish in nature [...] such distinctions; they are only necessary in the conception of the subject, in *memory, fear* or *hope*" (EPW 259 A). This is why Hegel said that, sooner or later, the meaning of all concepts consist in "vital and spiritual situations" (WL I 335).

The absolute lack of time that characterizes the time of physics, instead of diminishing, has been accentuated with the theory of relativity: the expression 'space-time' denotes this, for the four-dimensional space of Minkowski is simply space (not time), but physics have the whim of calling it so. It happens just what we saw (II 5) with decretory definitions: what they do is to stick to something different and leave time aside.

Eddington expresses this naively: "In the four-dimensional world we shall accordingly regard 1 second as the equivalent of 300,000 kilometres, and measure lengths and times in seconds and kilometres indiscriminately" (1978, 46).

Their procedure is very simple: since a six inches bomb irrigates one hectare of wheat, they take 'wheat' as a synonym of 'water' indiscriminately.

A more lucid physic, Herbert Dingle, has expressed this clearly:

> It is therefore because of our voluntary choice of measuring processes that we can speak of the 'union' of space and time into 'space-time'. This has nothing to do with any philosophical notions of space and time in themselves; it is a consequence of the wisdom of making sure that, after expressing motion in terms of space and time, the measurement of time shall depend on the measurement of space. Only on this account is it possible, as Minkowski did, to represent the time-co-ordinate as a fourth space-co-ordinate and describe motion geometrically as a track in a four- dimensional continuum. (Schilpp, ed., II 1970, 543s).

Although they intend to preserve relative time and evade absolute time, the physics of our century fail in fact to recognize time itself —something that also happened in classical physics with the postulation of an absolute time. This, of course, should not surprise us because time is not something physical.

Mathematical time is not time because —as Hegel demonstrated— it lacks true continuity. We will later study real continuity as real time. Following a very different path, physics introduced an 'infinite divisibility', which seemed to be a proper feature of space, but the *quantum* of Planck has demonstrated that it does not even exist in space.

According to the experimental discoveries of quantum physics, it is false that the material reality is infinitely divisible. If we start fractioning the material, we will find sooner or later something that is not further divisible. But it surprises me that physics had to wait for Planck, since Hegel demonstrated this in the text we quoted before (GP I 307): it is contradictory not only that time consists of temporal points, but also that space consists of spatial points. In addition, it is evident that something extended cannot be the result of unextended elements.

By means of a dilemma whose second branch derives into a new dilemma, it is easy to explain that, whoever affirms the infinite divisibility of space and rejects the *quanta*, forcefully affirms that the ulterior components of space are unextended points. The dilemma is the following one: the ulterior components are either divisible or indivisible. And if they are indivisible, they must necessarily be either extended (*quanta*) or unextended (points).

We cannot choose the first branch of the dilemma, for if they are divisible they cannot be the ulterior components; they can only be the resulting segments of the last division.

But if we say they are divisible, we would necessarily have to choose between extended or unextended ulterior components. Therefore, whoever rejects the quanta affirms that the ulterior components are points.

It is evident, however, that the sum of unextended elements cannot render something extended. Even before Planck's experiments, the only thing logically sustainable was that the ulterior components of space were *quanta* i.e. indivisible extended elements. Mathematical space —meaning a space that is infinitely divisible— is a contradiction of which Hegel timely warned us. Now, on the grounds of the spurious identification of space and time, physics always presented the image of an infinitely divisible time in substitution of a continuous time which is, as we will later see, the only true one. Physics incurred and keeps incurring in two big mistakes: first, to confuse time with space; second, to attribute an infinite divisibility to time that even space lacks.

It is of primal importance to notice that the confusion of space and time is a widespread epidemic: the layman does not have the concept of time. This does not wash away the blame of physics, since the purpose of science is not to reaffirm indefinitely the false beliefs of common people. Some figure that time is a fluid, as we saw Newton did; but a fluid is something spatial; evidently, they are imagining themselves being right at the side of a river, which is something spatial: but none of that can be comprised in the concept of time. Others —and this occurs more often— imagine a line that goes all across the imaginary visual field from one end to the other; they call the left segment 'past' and the right segment 'future'; they call 'present' a very little segment (or point, as they commonly refer to it) that falls within the visual axis.

All those things are spatial and have nothing to do with time. Besides, it is clear that all those images suffer from the aforementioned exteriority between their elements, and thus they lack the true continuity that characterizes time.

Again, what the physics and the laymen *have* is a fantastic image of time, not its concept. No image can be proclaimed the meaning of the *word* time because all images are spatial. They only need to *imagine* time to transform it into something which it is not.

The past imagined by the physics is of an illusory nature; it consists in a physical fact with its own observable characteristics, but the character of past as such is not empirically observable, not even by a magnifying glass. Even if it was situated in the past, the present or the future, the object would be entirely the same in terms of its empirical features. Therefore, time is an instance that the knowing subject introduces into the picture. He takes it not from the empirical data but from the spirit. A determined event belongs to the past *for the spirit*; and this is so because the spirit remembers it. Rothstein remarked correctly that it is difficult to give meaning to the word *past* without presupposing memory.

No object has a label that says 'I come from the past'. It is perfectly obvious that 'past' is something that has to be predicated from the present. Consequently, *the past has to be contained in the present*, and that proposition is not metaphorical but strictly rigorous. That is really a *continuum*; the multiplicity in unity; the multiplicity that is both identical with and different from the unity. It is not a *contiguum* or a juxtaposition of external elements. Just in the truly and authentic continuity can be time, but it only exists in the spirit. By no means can an event become from the 'past' only because I imagine it placed on the left side of my imaginary screen, or because I imagine it somewhere 'behind'. Both 'left' and 'behind' are spatial contents; they have absolutely nothing to do with time. If the past is not present in the present, it is no past at all, given that if it is not present, what we have is exteriority and not continuity.

Likewise, the present character of an object or a fact —in other words, its 'simultaneity with me'— is not empirically verifiable, even though physics have invested in this proof incalculable endeavors and witty arguments. First, it is obvious that, if the objective data was past or future instead of being simultaneous with me, its empirical features would be exactly the same; hence, its simultaneity with me is not one of its empirical features.

Second, having rejected absolute simultaneity for not being an empirical data, the theory of relativity tries in vain to preserve relative simultaneity —i.e. the simultaneity of two close events—, disregarding the fact that relative simultaneity is as unverifiable as the former. Actually, physics have to define sooner or later relative simultaneity like this: two events that occur in my proximity are simultaneous if I perceive them simultaneously. The *definiendum* reappears in the definition, and hence nothing has been defined and we still do not know what simultaneity means. Besides, it is obvious that this definition resorts to self-consciousness, for what it really says is the following: the events are simultaneous if my perception of the first event is simultaneous with my perception of the second event. This definition shows that empirical data cannot build up that dimension of time called simultaneity, just as it happened before with the past.

But there is still more to this. Whoever says that they are simultaneous if I perceive them simultaneously and in my proximity, presupposes necessarily that there is simultaneity between the close event and my perception of it, but that has two mayor drawbacks. The first one is circularity, for a simultaneity that is not being defined is presupposed, and that is exactly what we are trying to define. And the second one is that this same presupposition shows that simultaneity is empirically unverifiable. One tries to prove the simultaneity of two close events by the procedure of perceiving them simultaneously, but that very same procedure needs to *assume* that the occurring of the event and my perception of it are simultaneous due to their proximity. The simultaneity of two events locally close is not verifiable because every verification processes presupposes it.

It is worthy to notice that the *real* inclusion of the past in the present —which, as we said, is necessary for the past to be past— is also necessary in order for the present to be present. Bergson already stated that there is no self-consciousness without memory. That is one of the most valuable observations made on this subject. Consciousness of the present is not —as no act of consciousness can possibly be— an unextended atom of time. It does not occur in a timeless point, for that would not be consciousness of duration, i.e. consciousness of time. If it does not include the past, the present cannot be present. Those unextended instants just exist in the abstract thinking of mathematicians, not even in the imagination of mathematicians, since what is unextended is not imaginable; and, of course, those instants are not time, for they suppress continuity and duration, which are indispensable in order to speak about time. It is very symptomatic that experimental psychologists, being astonished and dazzled by the prestige of mathematics, have substituted the term 'psychological present' for the term 'specious present'. The psychological present has duration and, therefore, *really includes the past*; in contrast, by definition, the mathematical past does not exist, and this is the reason why a mentality strongly rooted in physics has to conclude that the psychological present deceives. They want to correct the only present that exists by means of an unreal one. Besides, they are violating the principle according to which nothing that is not subject to experimentation exists. The only present that exists is the psychological present, that is to say, the present of self-consciousness, since the empirical data do not suffice to build up the present, the past or the future.

We have incorporated the sincere recognition of Rothstein to our analysis by stating, without memory, the word *past* lacks meaning. It is absolutely transcendental to notice that, if memory or remembrances are considered simple reappearances of the experiences we had, we do not obtain thereof the idea of past, for such experiences did not present themselves as from the past because they did not have that feature and are only being repeated exactly as they occurred.

Such memory that can only repeat things is all what the materialist psychologists manage to conceive. But the idea of the past and the existence of time itself remain unexplained. True memory implies an 'I have experienced this', and consequently, self-consciousness, not as a simple 'companion' but as the identity of contents (cfr. III, 5). One requires the spirit, whose experiences have real continuity between themselves, because the spirit *consists* in its experiences and, as we saw, it is the substance whose accidents are not accidental.

Referring to experimental psychologists, Hegel said something that nowadays would be the solution to much of this discipline's problems. "The incapacity of understanding this universal which is in itself concrete but remains simple, is what has triggered by some people the belief of preserving peculiar experiences in peculiar *fibers* or *regions* (of the brain); according to them, the diverse can only have an isolated spatial existence" (EPW 453). What such psychologists do not manage to understand is that such localized and spatialized experiences would be just like the past of the physics, which could not be experienced as past, and hence is not past.

"The self must be a totality, and yet this totality must be as simple as the self" (EGP 277). The so-called big problem that psychologists face while dealing with time is not that big at all, since, on the one hand, the difficulty they come across with stems from the supposed intermediate periods of time during which the spirit does not act —which, by the way, are mistaken, for those periods of time are attributed externally to the soul—; on the other hand, that equals to attribute the soul something that is not time, for the simple reason that the time of physics does not actually exist. Capek says admirably: "...*forgetting rather than remembering* needs an explanation." (1971, 156).

As a complementary remark, let us just say that most of recent neurological researches do not provide favorable results to the so-called thinkers who believe in the storage of experiences in fibers or special regions. Every day we witness more cases in which the extirpation of a zone that was happily declared the monopolizer of certain function only has as its consequence the transitory numbness of the said zone. Ernest D. Gardner comments: "nevertheless, the anatomical distinctions are often not clear-cut, the connections between nerve cells are enormously complicated and not at all well understood, functional specificity is often lacking, and species differences in function are hardly understood at all." (EB 24, 832, 1)

Also, in the case of oblivion, emotional inhibitions and personal interferences originated in interpersonal relationships have undoubtedly a great bearing. We will soon see that the constituent of the self is interpersonality. The following observation of Benton Underwood has a reach that goes beyond biological adaptation:

> ...it is helpful to consider what would happen if memories failed to fade. [...] Without forgetting, adaptive ability would suffer; for example, learned behavior that might have been correct a decade ago may no longer be. Cases are recorded of people who (by ordinary standards) forgot so little that their everyday activities were full of confusion. Thus, forgetting seems to serve the survival of the individual and the species. (EB 23, 947, 1).

We do not even need to deny the 'enagrams' or the recording of experiences in certain areas of the nervous tissue. With or without them, what happens is that true memory remains as unexplained as it was before. At the most, the exact repetition of memories would stem from them; but that means, as we previously said, that they do not appear as past memories, for the simple reason that they did not have that character before and, in theory, they are being repeated exactly as they occurred.

Without an 'I already lived this', there is no past and, consequently, no time. The self identifies itself with its experiences; it is not a substratum different from them; on the contrary, they build up continuity in the strict sense of this word: "the continuity of a universal and a being-with-itself which is not interrupted by the infinitely different limits which are constituted by the impressions and intuitions" (WL I 182). We have seen how in the content of the self all other contents (v.g. substance, being) are included, without this implying that the self ceases to be what it is or that it is interrupted. True memory is of a conceptual nature and implies thought and congruency: the imaginary is only a sequel. And Hegel says about this sequel:

"The proceeding of intelligence in the imaginations is both to interiorize the immediate and to put itself as perceiving in itself, and to suppress the subjectivity of interiority by freeing itself from itself and from the being in itself in its own exteriority" (EPW, 451).

With respect to the material that comes from empirical impressions, it is obvious that "only in thought does it turn into the concrete immanence which is the concept" (ibid); only on that level are we aware of it, and, therefore, when it comes to remembering it, that level is the most decisive one.

Therefore, there should not be any misunderstanding about the term memory, when we say that physics strive in vain to make the word time to have a physical meaning, i.e. to make it to have a meaning without the intervention of memory, the self and self-consciousness. They mention, for instance, the lapse that goes between two sunsets. Besides the fact that the words 'lapse' and 'goes' lack empirical meaning, (cfr. II, 5) human intelligence has reached the necessary maturity to face, in all its harshness, *the fact that two successive sunsets is not an empirical data*. Furthermore, even if these sunsets were exactly the same in terms of their empirical features, we could say that *their being two* is not an empirical data. Sensibility cannot perceive numbers. Empirically speaking, each of these dawns is what it is; they exist independently from each other and have nothing to do between them. Only memory makes them successive, and only there lies time. The time of the physics is not time.

After joyfully rejecting absolute time for being empirically unverifiable, it is unsettling that physics do not realize that every kind of time is empirically unverifiable, not only the absolute one. They even seem to incur again in the postulation of absolute time when they dump at us calculations of millions of years, not only forgetting that the rotation of the Earth is not regular and that, in principle, the regularity of the vibrations of cadmium is unverifiable, but also that, if all cosmic movements were ten thousand times faster as they are now —or, in any case, ten thousand times slower—, we would not notice absolutely anything, and it follows that those mind-blowing numbers lack all physical meaning just as absolute time does. Besides, the impressiveness of those numbers depends on the mistaken presupposition of time as a strictly contradictory continuity that does not even exist in space (cfr. supra). The public can peacefully remain calmed before such staggering numbers: the time of physics has nothing to do with real time.

What we said about 'successiveness' can be said of what is 'periodical', for the former enters in the definition of the latter, and that makes it something not empirical. We present this warning because of those careless scientists who believe they can define time by the hands of a watch and avoid Einsteinian circularity (cfr. II, 5) by saying: that a watch is an 'isolated system that runs periodically'. Perhaps circularity is avoided, but one does thereby without empiricity as well, for without the intervention of memory and self-consciousness it is impossible to provide meaning to the word *periodicity*.

7. Intersubjectivity

We will now proceed to make explicit the third characterization of spirit, which was lying underneath the two former ones from the very start: intersubjectivity —which is, essentially, ethicity.

If, as we have seen, the subject *consists* in the awareness of the self, in that mental and ideal content called self-consciousness, then intersubjectivity is essentially there from the very beginning, for it is obvious that a child does not manage to call himself a 'I', he does not gain awareness of his self —and thus, he is unable to create it for the first time— until his mother (or somebody else) addresses him in such a way that, by personalizing him, he comes to realize that he is a person; analogously, by giving him responsibilities, he gradually becomes a responsible subject. In the awareness of the self that is being formed the awareness of the other is present in such a way that they are not only inseparable: they stand, in fact, in such a degree of mutual penetration that the child has awareness of himself by having awareness of other, and this intersubjective constitution of self-consciousness is always constant, because the consciousness of one's own identity is consciousness of one's own distinction with respect to others, and hence it is consciousness of others. Therefore, self-consciousness (i.e. the self) and the consciousness I have others nourish one another; they make each other exist, and, truly, one is the reason of being of the other. It follows from this that his third characterization does not pretend to point out an extrinsic cause of the self but rather its very being and its intrinsic constituent. Now, this consciousness of the others —which makes my self-consciousness— is of an eminently ethical character. Its content is the following: *I do not exist alone; the others also exist*. The others are subjects —not objects— just like I am one; I cannot confuse them with things that can be treated as means, because they are self-consciousnesses just as me.

If this was not the content of the consciousness I have of the others, if it did not make me responsible, if it was not an exigency that guides my freedom by constituting it, it would not generate self-determination and, therefore, it would not cause a self, for we saw before (III, 2) that, essentially, the self makes itself what it is. To this exigency or imperative the self can answer in very different ways, in very different degrees of consent, in very different degrees of responsibility, but by answering to that exigency it provides itself with its own determinations. Furthermore, it conforms right from the start the mass of which moral substance is constituted --but that, of course, is only a metaphor, since we have seen that there is not a substratum. The self is not a material rock: its only real consistency is the mental and ideal content called self-consciousness, as well as the consciousness I have not only of my existence but of the existence of the others —which, most certainly, is the very moral content.

Let us not forget what we previously discussed (III 5). We said, following Kant, that there cannot be consciousness without self-consciousness. And we just added that there cannot be conscience without self-conscience. This is one of the most penetrating discoveries; it is dated X, b.C. —as we will later see (VI, 1)— and was made by an anonymous thinker who is called the Yahvist by the exegetes. Hegel takes his discovery extremely seriously.

The Hegelian formulations of our third characterization of spirit enrich and complement our exposition. This thesis plays a fundamental role in the philosophy of Hegel, as it has been noticed by both his smartest (Laure, Mure, Stace), and his superficial (Hyppolite, Kojève, etcetera) commentators.

"Without a thou, the self is impossible" (JS 378).

"Spirit is essentially being for the spirit, and it is only spirit in so far in that it is for the spirit" (PR I 201).

"The true contrast that the spirit can have is spirit; only by means of its strangeness in itself can it reach the force of being spirit" (WG 535).

The otherness refers to the 'you', the other. Otherness in itself means the consciousness-of-the-other that I have in myself. According to the *Philosophy of History*, only through it the spirit acquires the capacity of having a self and be properly a spirit. "the subject must turn to another subject" (PR III 133).

"Self – consciousness exists in and for itself when, and by the fact that, it so exists for another; that is, it exists only in being acknowledged." (PG 141)

"As consciousness, each is for the other the middle term, through which each mediates itself with itself and unites with itself; and each is for itself, and for the other, and immediate being on its own account which at the same time is such only through this mediation." (PG 143)

"Only in moral this concept of absolute individuality of consciousness exists and makes everything" (GP I 271).

"This ethical substance which constitutes the spirit, forms the life and the essence itself of individuality" (GP II 108).

"The ethical, the laws of freedoms, is the supreme and purest form of the spirit; according to its nature it is not something spiritual exterior to it; it is not something extrinsic or fortuitous, but only the nature of the spirit itself" (PR III 19s).

"That the right being in itself and being for itself, is what I am in the ethical substance; this is therefore the essence of selfconsciousness". (PG 312)

The exigency, the imperative —which we stressed in the consciousness that the self has of the others— imposes to the subject in embryo —the child that is turning to self-consciousness— the moral necessity of suppressing his merely natural instincts, his animal instincts, for not only he but also the other self-consciousnesses exist in the world, and he cannot treat them as means: this is how a child ceases to be an animal and becomes a spirit. "The soul is spirit only through suppression of the natural willing, of the appetites. That happens when one is subject to the ethical, that is to say, by becoming accustomed to the ethical, so that it becomes the second nature of the individual" (PR II,178).

"Man is addressed with the exigency of not being as a natural will, of not being as what he by nature is". (PR III 107).

"The natural is rather what the spirit has to suppress" (GP II 107), "the natural man is egoist" (PR III 115s), "the naturality of the will is precisely the egoism of will" (PR III 116).

The essentially ethical and intersubjective character of the spirit is one of the fundamental theses of the phenomenology. One only needs to take a look at the first heading (A) of chapter six where Hegel deals *explicitly* with the subject that gives its title to the work: "the true spirit, the ethicity". Hegel says there: "The true spirit is the real and absolute being that holds itself together" (PG 314). As Lauer comments (1977, 92), this is the reason why the work was first called *Science of the Experience of Consciousness,* (cfr. PG 61) and under that title it was announced in the bibliographical bulletins of its time, but with its ulterior development it finally found its title in *Phenomenology of Spirit*.

Let us make a brief hermeneutical excursus. The aforementioned suppression of the instinctive egoism for having to work for all, the necessary repression of natural impulses and appetites (*Begierde*), is what makes the spirit of the slave more capable than the master of becoming a true spirit, without forgetting that this dialectics (PG 141- 150) belongs to the past figures which, as Hegel warns us, are mere abstractions that do not lead us to ethicity. In the chapters of the Encyclopedia that thoughtfully summarize this dialectic we find this explicit remark: "By serving his master the slave gradually demolishes his own privative wants, suppressing the interior immediateness of his appetites" (EPW 435). In the *Phenomenology* itself, it is affirmed that in serving the conscience of the slave "suppresses in all its particular elements his attachment to the natural existence and he rules it out" (PG 148). The *Encyclopedia* says in more detail how this happens: "by means of the negation of immediacy, that is to say, the appetites" (EPW 434). The ethical character of this situation is analyzed in the *Philosophy of Right*, a propos of the institution of the family: "the egoism of the appetites transforms itself into the care and the acquisition for a *collective*, in something ethical". (RPH 170) Therefore, in order to affirm that work itself in its materialness is what, according to Hegel, transforms the slave, (así *Marx en México*, EB I 574 y Kojève 1947 passim) one needs to omit deliberately the key passages and substitute them with the own thoughts of the reader. End of the excursus.

The thesis according to which self-consciousness stems from intersubjectivity is not only obvious for what we know of child psychology. Anthropologists and sociologists have studied in detail the efficacy —in the formation of the individuals' selfhood— of the 'roles' and the corresponding 'expectations' in a society. Only an immoral tendency could have sometimes prevented one from recognizing that the most basic expectations and the most fundamental roles are of a moral nature.

Let us consider this question: are there roles and expectations without which any kind of society or human community is not possible? For instance, I have the expectation of not being killed by those who come close to me for any given reason; I have the expectation of hearing the truth from those who talk to me; I expect that the others keep the promises they make, etcetera. These expectations impose the respective roles, and all persons bear the same expectations in regard of me and impose me different roles. Without that no human community is possible. When a child incorporates to the preexisting intersubjectivity, these roles and expectations shape and penetrate his subjectivity —and the moral character of the said roles and expectations is undisputable. The more free-spirited anthropologists have showed this by referring to the most primitive human communities nowadays known. For instance, Robert Redfield says:

> The point in which we must insist [...] is that the nexus and the essential order of society in such an early condition of humanity were moral. Humanity reached its enduring and characteristic nature as a multitude of different yet equivalent systems of relations and institutions, each of which was expression of their conception of the good. Each precivilized society was bound by certain ethical conceptions which were most of the time implicit but continuously carried out. (Rossi et al., 1977, 291)

As we shall see later (VI, 2), the utilitarian dogmatism which interprets the individual search for self interest as the moral precepts without whose observance no society can exist, lacks significance and refutes itself.

8. Causality

Two objections had appeared against the characterization of the subject that we previously presented in three steps: one objection is based on the concept of substance and the other on the concept of time. The last objection —which we will answer here— is supposedly based on the concept of causality and emerges directly against self-determination, but since it is a frustrated attempt, it rests on the idea of natural law and physical necessity. This is the reason why we have postponed this discussion for the very moment in which we have reached the moral necessity called imperative. As we will later see, physical necessity lacks all meaning whatsoever, and the origin of the concept of necessity is precisely the moral imperative.

Against the characterization of the subject as pure act, the two former objections put forward the ideas of substance and time, only to collide against the fact that the only possible meaning that these two words can have is the activity itself of the spirit which they are attacking. Now, the same happens to those who put forward causality and necessity against self-determination: the only possible meaning of these words is constituted precisely by the reality they want to raise up against. The objection says that there cannot be self-determined beings since science holds as a fundamental principle that all phenomena are caused and determined by other phenomena; but the meaning of the verb 'to cause' remains here unexplained, unless one recurs to self-determination, which means that one causes oneself his own determinations.

No one would be surprised today by what we just said if scientists had not taken refuge in their own conceptions, exempting themselves of knowing the contributions that the most piercing minds have rendered demonstratively to humanity. Indeed, Hume and Kant demonstrated once and for all that the concept of cause neither does nor can have an empirical meaning. Hume mocked all those who empirically perceive (or believe to perceive) a *post* and recklessly affirm a *propter*. There is so little logic in that recurrent ingenuity as is in someone who, at the sight of an ant, would say that such thing is an elephant. If the succeeding from one event to the other meant that there is a causal relationship between the two, we would have to affirm that the night is the cause of the day, because the event called day regularly succeeds the event called night.

Hume says: "Shall we then rest contented with these two relations of contiguity and succession, as affording a complete idea of causation? By no means. An object may be contiguous and prior to another, without being considered as its cause." (*Treatise of Human Nature*, I, III, xiv)

Kant simply is analyzing the undeniable content of the concept of cause when he affirms that it is impossible to grant this concept an empirical meaning:

> Sensible phenomena provide certainly cases by means of which a rule is possible according to which something happens habitually, but never that such a result is *necessary:* it follows that the synthesis of cause and effect has a dignity that cannot be expressed empirically, namely, that the effect not only is added to the cause, but rather is put *by* and follows *from* it (KRV B124).

It is important to repeat in this context a warning that we made before (II, 3). Although it may be valid in some case to *infer* a causal relation on the grounds of empirical data, it does not follow from this that what is inferred is an empirical data. Quite the contrary: it is necessary to make this inference, because it was not among the empirical data. By the way, the meaning of the concept whose realization we infer has to be taken from somewhere else, since this meaning does not match any of the empirical data available.

In any case, the fact that B is contiguous to A or that it occurs after A *does not mean* that B is an effect of A. Contiguity and causality are not synonyms. Temporal succession and causality are not synonyms. Replacing surreptitiously one for the other is always a sophistic maneuver

To add B to A could be an empirical data. From that one could *infer* that B is an effect of A if he finds the corresponding thesis and demonstrates them. But that does not mean that being an effect is an empirical data. On the contrary, the proof that this is not the case is that it was necessary to infer *that*. An empirical data does not need to be inferred, since it is already given. By the by, these inferences would be pointless pursuits if one does not define the meaning of 'effect', since we would not understand what has been inferred.

But there is something worse: there are no adequate premises, the inference is invalid. Although it were empirical —as unreflective scientists assume— that B is *always* added to A, it is not possible to build a legitimate inference by means of which we deduce a causal relation between A and B. They will reply: it will be too much chance that there was no causal relation. However, I answer them: precisely this reasoning shows that they are using an entire theory to legitimate the inference and that the causal relation is not an empirical data, but that theory is false, because the night *always* is added to the day and nevertheless there is no causal relation between the day and the night, and it is useless to say that this constant succession would be too hazardous if there was no causal relation; this does not exist, despite all the theories of causality they may come up with. In addition, an 'always' is not an empirical data.

And it is also pointless that they employ the recourse of saying 'we understand by causal relation a constant succession'. It is simply false that they *understand* that, because they *do not understand* that the day is not the cause of the night.

What we have said shows that causality is not an empirical data —something that was already demonstrated by Hume and Kant. Now, it follows from this that only self-consciousness can give meaning to the concept of cause. But the only cause we know by self-consciousness is the cause that determines itself. Therefore, one cannot put forward against determination a concept whose only meaning is self-determination.

A disturbing but logically unavoidable conclusion is that if we attribute causality to other kinds of processes and realities, we only do this in a derived, deficient and diminished way. And even this softened attribution is probably not entirely justified.

Since they did not venture to read Hegel, both the enemies and the champions of free will believed —as an undisputable fact— that free will is incompatible with causal processes. The first ones denied the existence of free will because they thought that the entire universe consisted in causal processes; the second ones affirmed it by saying that there are free processes in the world besides causal ones. But all the controversy rested on a concept of cause that does not exist. The only true cause is that which determines itself, i.e. the free will.

No one in fact thinks something defined when one uses the word 'cause' in reference to material phenomena, not even in the case in which a moving body collides with another and the latter starts to move. One says 'cause' as if it were some sort of 'hocus pocus'. When vulgar or professional philosophers and scientists believe to have explained an event only by saying that it was 'caused' by some other event, they are really explaining *obscurum per obscuris*. What could it possibly mean that a being 'makes being' another being? How is it possible that the property or characteristic of a being 'transmigrates' and suddenly becomes the property or the characteristic of another object? What kind of metaphysical monster is that which was first an accident, then a substance for a moment of transmigration migration, and finally becomes an accident of another substance?

"The relation called influx belongs to vulgar philosophy. Since one cannot understand how material particles and immaterial qualities can pass from one substance to the other, one must abandon such a representation" (GP III 240).

The evidently mysterious and magical character of the transmigrations mentioned above did not prevent Hobbes and Huygens in the 17th century, Lord Kevin in the 19th century, and the molecular biologists of our century, from playing with billiard balls, as if the movement of the first could explain the movement of the second only because there was 'contact' between them. Furthermore, this illusion of the explanatory force of something that is in fact more mysterious than that which was being explained underlies the grounding of the entire corpuscular theory of matter, which tried to reduce everything to the clashes between marbles. And all the present efforts from the General Relativity onwards of abolishing Newtonian actions by substituting them with different kinds of 'contact' are still trapped within the disastrous Kelvinian framework; meanwhile, the scientists stubbornly persist to explain the obscure with the darker.

If one replies that the property which first was in one being and then in another is not numerically the same, the magical character and the lack of all explanation become all the more clear: from the fact that a being has some characteristics one invalidly proceeds to say that the *other* being has such characteristics only by mentioning the word 'cause'. How does this marvel occur is something that remains unexplained; one thinks it is enough to say that one is the cause of the other.

And if it is numerically the same property, the circumstance of the contact does not make more comprehensible the mysterious fact that the characteristic of an object —despite the fact that we are talking of the individual entity itself of that object — becomes a property and the individual entity itself of an object that is different from the former one. We must remember: contiguity is not a synonym of causality. Hume had already remarked this: "Motion in the second billiard-ball is a quite distinct event from motion in the first; nor is there anything in the one to suggest the smallest hint of the other." (*Enquiry* IV, I)

Despite the degree of contiguity of two beings, everything remains unexplained unless one says how one acts upon the other. Therefore, the contact itself is not decisive. And if it does act, then that causality would be what builds up the explanation. For that purpose, contact itself is completely irrelevant.

But precisely this term of causation lacks all intelligible meaning without the reference to self-determination; as we saw with Kant (III 2), self-determination is the most intelligible of all the contents that the mind can understand. What happens to a billiard ball is by no means explained by reference to what happens in *some other* ball. To say that what happens in a material object is the cause of what happens in another object equals to pronounce a spell without providing any intelligible content.

The root of this widespread Kelvinian mistake is twofold: first, one believes that what is frequent (everything that simply occurs), does not need an explanation; second, one incurs completely in the sophism denounced by Hume: *post hoc, ergo propter hoc*.

Since the ball always moves after a contact, one believes that this fact, only because it always occurs, does not need an explanation. But true science is characterized by the search of the most ordinary and common facts, that is to say, the facts that always occur. For instance, the theory of gravitation tries to explain the fact that a stone falls to the ground when we drop it. Likewise, the tides are always shifting, but only geology tries strenuously to explain that movement. We already pointed out that if we ask 'why are there black clouds when it thunders?', and one answers us that 'every time that thunders there are black clouds', no one has explained anything.

But what is decisive is that the Kelvinian mentality incurs in the tremendous sophism denounced by Hume. Since the second ball moves *after* the contact, the Kelvinian mind affirms: it moves due to the contact. We do not need to repeat that such procedure is as illegitimate as saying that an ant is an elephant, nor that with such a justification we would have to affirm that the night is cause of the day, nor that it is pointless to say that, if the second ball moves after the contact, it would be too much coincidence that this was not provoked by the contact; for we could also say that, if the day succeeds the night, it would be too much coincidence that this did not happen due to the night.

As if the previous examples were not enough, modern physics has provided us with another argument against the explanation by contact: the contact does not exist. This is a thesis worth looking at more closely.

The fields of repulsive force prevent that two bodies touch between them, and this happens even when the vulgar mind believes to be witnessing a clash. I quote the mechanical engineer Malcolm McChesney:

> In any real gas, even a binary collision —one between two particles only— is a complicated process, because each particle is surrounded by a force field that varies in space. This force field is such that it attracts another particle when that other particle is relatively far away but repels it when the other particle comes relatively close. That these intermolecular forces exist is evident from the fact that gases do condense into liquids, a phenomenon that cannot be explained except as a consequence of attractive forces at work between the particles. (EB 23, 697, 2)

McChesney even says this in regard of the explanations of clashes between bodkins:

"The particle description of gases grew from ideas in the 17th and 18th centuries concerning special aspects of the collective behavior of structureless gas particles that were supposed to have no surrounding force fields." (EB 23, 699, 1)

In Physics one cannot speak of contact anymore, except only when this consists of the forces of a body that act on the other, i.e. one exerts causality upon the other. But then, something very disappointing happens: they wanted to explain causality by means of contact, when in fact contact is explained by causality. One employed supposedly a concept in order to make causality intelligible, when in fact the only possible contact would be intelligible if causality was intelligible. This is a bold circularity that confirms what we have been saying; one does not think in something definite when one uses the word *cause*.

There is another conclusion in regard to the efforts of General Relativity mentioned before —and the more recent ones, based on spaces of eleven dimensions— which aim to overcome the Newtonian action at distance. If they conceive distance as the negation of contact, that equals the negation of causality, in other words, the negation of the action of a body upon another body. Therefore, the action at a distance turns out to be a strictly contradictory term that lacks content, for that content would be action without action. What the relativists are challenging is nothing more than a ghost.

The true problem is not —as the Scholastic philosophers also believed— the action at a distance, but rather action in itself and causality. And in that regard, physics have nothing substantial to say, because it is impossible to define causality on the physical level. The only causality that we know is the self-determination of the spirit, which really produces new experiences and determinations that did not previously exist. To use the term causality with this point of reference is only to be juggling with words.

For that reason, spirit is the only possible meaning of the word time, since as we saw in Aristotle, there cannot be time where there is no real change, i.e., the production of something new, and this is why Physics —which lacks the concept of causality— had no other option than to misunderstand time, confusing it with space.

One could vaguely foresee this when Einstein tried to reduce time to causation. His unexpressed principle was: *Propter hoc, ergo post hoc*. But since in Einstein —as in all Physics— causation is not really causation or production of something new, this utterance remained barren. As we saw, the time of Einstein is even reversible because it is nothing more than space. That Einstein does not understand by causation the production of something new is seen in his quest for some sort of contact, suppressing distance by the curvature of space and the false mathematical continuity of the field. One must say emphatically that the Kelvinian explanation by contact presupposes that nothing new is produced: the movement of the first billiard ball is simply transmitted to the second one. Parmenides ho! *Nihil fit, nihil movetur*!

All the 'explanations' of Physics converge in some conservation principle: everything is conserved; there is nothing new.

As we saw, the subsequent mistake is that —following Parmenides— they have confounded being with nothingness. Nothingness, indeed, does not change. It neither increases nor diminishes. On the contrary, the true being, which is the constant self-determination of the spirit, consists in producing genuinely new determinations because it only exists in doing that.

This is why Aristotle said that the atomists lacked the 'origin of movement'. As many commentators have noted, the atoms of Leucippus and Democritus aim primarily to be the being of Parmenides in miniature, which is tantamount to say that nothing changes. Now, the same thing has happened to all physics. All what the Parmenidian 'explanations' by conservation —that is to say, by identity— do is to deny that there is something to explain: nothing changes, there is nothing new. Thus there is no time or causality which can be properly called as such. Physics can use the word causality; since they do not need the real causality, they do not give this word any meaning at all.

In balance, thus, causality without content cannot help to challenge self-determination; whether it has content or not, this causality not only does not challenge self-determination: they are identical.

9. Natural law

Consequently, in order to keep denying the concept of self-determination, what some scientists have done is to renounce *de facto* to the concept of causality and instead they brandish the notion of natural laws.

All modern sciences entrench themselves in the caste of natural law; all the efforts aim 'to discover' this law, and they lay in that the entire scientificity of their disciplines. But they do not manage to define by empirical means what a law or a nomological proposition is, and hence no one knows what they are taking about. Now, if the meaning of this word is not empirical, then it must come from self-consciousness, which leads us directly to the moral imperative described by our third characterization of spirit, and is precisely what the empirical disciplines wanted to avoid.

To tell the truth, it is a perfectly documented fact that the word law —I refer, evidently, to ancient tongues— was employed in a moral and juridical sense a millennium before it was being used by Physics. The code of Hammurabi, which dates back to the 18TH century b.C., says in its introductory paragraph that the law chisels the stony walls of the royal palace so that all the population has knowledge of it. What is really questionable is whether the word *law* can obtain a meaning that is independent from its original moral and juridical sense. For that to be possible, it would be necessary that an empirical data builds up a new meaning. However, none of the keywords that must be employed to define or enounce a law has an empirical meaning.

It is amazing that a word, whose original meaning is the imperative directed toward rational beings capable of being responsible, was ever someday used in reference to irrational things, beings and phenomena.

Evidently, this second use had to be derived and with a deficient meaning, as it happened with the word *causality*. Now, transfer or apparent moving away from its original sense is real and is documented; it was a fact of a clearly theological nature and its result does not have the slightest empirical trace. It happened when humanity thought that God gave laws to the entire universe, not only to rational beings; this moment is documented in the Hebrew word *huq*, because it is undoubtedly a very common idea among the ancient tribes. We read in the Bible: "When he gave to the sea his decree, that the waters should not pass his commandment: when he appointed the foundations of the earth" (*Proverbs*, 8, 29). The same meaning can be corroborated in Jeremiah, 5, 24; 31, 35; 33, 25; Job 38, 33.

Without this theology it would be an unexplainable historical fact that at some point one started to speak of laws or imperatives with reference to material things; from this comes undoubtedly the *use* of the word *law* that is nowadays common in sciences, although the scientists would desire that its *meaning* were different; the fulfillment of this wish would only happen once they can give the word law an empirical meaning. They will never be able to do this.

But the above mentioned documented historical fact has also repercussions of huge transcendence backwards; it sheds intense light on the nature of the original meaning itself of the law as an imperative that regulated interhuman relationships and that made society possible (cfr. III, 7) for the first time; this fact shows that this imperative —constitutive of the self and self-consciousness— was perceived as the law and voice of divinity. As the *Phenomenology* says: "It is the moral genius which knows the inner voice of what it immediately knows to be a divine voice" (PG 460) This is why Hegel says: "Religion is the first way of self-consciousness" (VG 125).

Any anthropologist knows that, in the origins of mankind, "the civil and state laws were completely identical to the religious laws" (PR I 270).

"We know from God's existence, and this knowledge is present in us in a way so immediate that it becomes authority, the intrinsic authority of conscience" (EGP 195).

"It is certainly true that men must be educated towards religion, but not towards that which is not there yet" (VG 128).

If the original meaning of the word *law* had nothing to do with God, it would have been impossible that somebody else thought that material things also have law.

The Anthropology of our century corroborates widely what Hegel just told us. Summing up numerous field investigations, Paul Schebesta affirms: "Up to now it has not been possible to find a people without religion" (König, I, 1960, 599). In the same line, Wilhelm Koppers says: "There is a lack of the remembrances and signs of an atheistic or pre-religious state of mankind" (*ibid.* 130).

Likewise, Ino Rossi makes his the result of Redfield investigations: "...in preliterate societies people are united by 'moral ties', that is, by the fact that they share the same moral and religious principles and that these principles are the source of motivation for daily behavior" (Rossi, 1977, 243s). On this account, Redfield himself adduces the words of atheist anthropologist Arthur Kroeber: "The members of these societies 'believe in the sacred things'" (*ibid.* 291). It is well known that even Comte recognizes that the first stage of humanity was theological.

One should notice that what is here at stake is something more than the 'interesting' and subjective folkloric way in which the primitive man perceived the imperative that made him being what he is. Paleonthologists have demonstrated that what is properly human does not date back that long ago if one considers integrally the history of life in this planet; in order to do without some problems, science cannot go back to a past of the human race which is not limited. Since we saw (III 7) that it is impossible that man exists if somebody else does not address him —because "without a thou the I cannot exist" (JS 378), it would be dogmatism not to ask who made the *first* man by addressing him. The atheist Leslie White plays with the text of Saint John by saying 'in the beginning was the Word". But this is a serious matter. It is comical to declare as Kroeber does (1969, 2) that science is not interested in the origins, when all modern anthropology is inspired on a book titled *The Origin of Species*. Leslie White himself affirms the scientific character of the following question:

> If mathematical ideas enter the mind of the individual mathematician from the outside, from the stream of culture into which he was born and reared, the question arises, where did culture in general, and mathematical culture in particular, come from in the first place? How did it arise and acquire its content? (1964, 278)

Let us move forward. We said that some scientists remain attached to the natural laws and to their inherent necessity of denying the self-determination of the subject, but they do not manage to clarify what they are saying because none of the keywords required to define or to enounce a law have an empirical meaning. I have sufficiently demonstrated this in *Appeal to Reason*, but it seems inevitable to summarize this in a very concise way.

An example: every acid reacts with a base and forms salt. It is obvious that the word 'every' goes beyond empirical data. We can only perceive *some* acids: who knows what the meaning of the term 'every' might be! The question here is not to determine whether it is legitimate to *infer* from some particular cases that all acid behaves in the above mentioned way; this is other problem, the problem of induction, and Hume already demonstrated that the inference is logically unsustainable. What I say is the following: even if we were to suppose that the inference was legitimate, the meaning itself of the word 'every' does not coincide with any of the empirical data nor with the whole of the really empirical data; this is why it is would be necessary to infer it, because we do no experience it. Although they might argue that 'in principle' all cases are perceptible, this statement would lack sense, since they have not defined 'every'. Instead of an empirical data, what they are delivering us is a statement that is not understood. Furthermore: instead of an empirical data they are delivering us a *supposition*, and a supposition is in essence unverifiable and, most probably, false. Since the acts of observation would be carried out in the future, that supposition would imply this principle: 'The future is similar to the present', which is unverifiable and most probably not true. It supposes that the conditions of observation would not be modified in the future, that the sensorial organs will not be modified, etcetera. None of these speculations gives the word 'every' an empirical meaning.

Another example: always that an acid is added a base it makes a reaction and forms salt. It is evident that no empirical data matches the word 'always'. What we can experiment is not an 'always' but some or many particular instants; the always is not sensible; even if we dedicated our lives to experience particular instants, we would never have the experience of an 'always'.

Another example: law is a presupposition that describes the constant way in which some things behave. The reader is able to realize that the word 'constant' could not be defined without the intervention of the word 'always' or the word 'all'.

Another example: the only behavior that can have an acid if one adds a base to it is to make a reaction and form salt. It is obvious that the word 'only' does not have an empirical meaning. If there is only one object in the visual field, the data is that object; the idea that there *could be* others but that we only have this one is a speculation of the intellect based entirely on unreal conditions; those other *possible* objects are not empirical data, precisely because they are not present, but the present object could only be called unique in contrast to the non-present ones. To say that this objet is unique amounts to saying that other objects are *not* present, but a 'no' is not an empirical data: in order to interpret the vacuum as absence or as negation of something, one requires an act of the mind that denies, because senses are not able to deny. If the senses deliver me something in blank, that blank is not a negation of anything whatsoever; it is simply blank; it could turn into a negation only if one added a consideration of the intellect that compared that vacuum with some of the objects that are not empirical data because they are not present.

Another example: if one adds a base to an acid, it reacts with it and makes salt. It is clear that no empirical data can be qualified as the meaning of the conditional particle 'if'. One cannot smell, see, or touch an 'if'. It is reflection the one that brings up the 'if'; therefore, it is not an empirical data but something *intelligible,* although it is legitimate that intelligence introduces it. As I have pointed out before, it would be an inference. With more virulence that other logical particles, the conditional particle 'if' is particularly indefinable by means of empirical data, for all the definitions carried out by logical atomism employ it; i.e. the make it appear on the definer; that is why we would have circularity in our case and, in the last instance, nothing would be defined. The attempt of atomism could be defined thus: the proposition 'if A then B' means that B cannot exist *if* A exists. The reader can already see that the 'if' reappears in the definer and that the circularity is shameless.

Another example, an acid cannot react with a base and form salt. We said that 'no' is not an empirical data. If there is an object within the visual field, this object is what it is and it does not constitute the negation of anything whatsoever; if there is nothing in the visual field, there is no empirical data either; there is not sensation at that moment; therefore, it is false that sensation perceives the 'no', for that cannot be perceived.

In addition, the nomologic formulation that we are considering contains the word 'can'. But no sensible data deserves to be called 'can'. At best, the senses witness that something is this or that way; they do not say *that they can be* this or that way. If it pleases us, we will *infer* from what they are what they can be (*ex facto ad posse valet illation*), but this demonstrates that 'can' is not being perceived empirically, because it needs to be inferred. For the inference in question one requires a premise that says: everything that is can be. In this premise the 'can' appears for the first time, and does not correspond to any empirical data; it is rather the discursive intelligence the one that introduces it. I do not doubt that it introduces it legitimately; that is another question. But it is something intelligible, not something sensible, for it was not in the mere sensation; it only appeared for the first time in the premise brought up by intelligence.

The 'no' and the 'can', each on their own, are enough to turn unempirical the formulation of the law that we are dealing with. But if they are brought together in a 'cannot' the unempiricity becomes sordid. In that case, not even the inference is valid, not to mention the experience. From an 'it is not' no one can extract an 'it cannot be' (*ex non facto ad non posse non valet illation*), and the senses do not even deliver us the 'it is not', since as we already see, 'no' is not an empirical data.

It would be superfluous to stop in naïve formulations of the law, which boast they do not need an 'all' and say: an acid reacts with a base forming salt. It is obvious that if the 'an' has the intention of a singular, the formulation in question is not a law; and if it has a universal intention, the formulators wanted us to understand it as an 'all' and their attempt of concealing this was in vain.

To tell the truth, we do not need to lengthen our journey. The example that we have mentioned shows that, if a formulation of the law really becomes a law, it would necessarily be the logical equivalent of any other formulation of the law. Now, if we have demonstrated that one of them is unempirical, that also applies to the others, because they are the logical equivalent of the first one. This is why all the attempts in the future of formulating empirically a law are doomed to failure. The reader sees immediately that 'an acid cannot' means the same as 'always that acid…' or that 'every acid…' etcetera.

10. Necessary

The only formulation that deserves a closer look is this one: 'an acid necessarily reacts with a base forming salt'.

The reader notices immediately that it is the logical equivalent of 'an acid cannot no…' Therefore, it is as unempirical as this and all the former ones.

A symptom of intellectual despair has been the critical acclaim that Saul Kripke's attempt at defining 'necessary' and 'necessarily' obtained. It first defines 'possible', for he thinks that what all possible worlds have in common will be a good definition of 'necessary'. He says that an M world is possible if it is logically compatible with the natural laws that govern our real natural world M. Here again the 'if' particle reappears —a particle which we have seen is unempirical. That very thing frustrates the whole enterprise. Afterwards, Kripke considers all possible worlds M1, M2, M3… and triumphantly exclaims: what they have as a common denominator is the content of the word 'necessary'.

This illustrious theorem has so many deficiencies that one can hardly believe that it has been taken seriously at all. We have already pointed out its gratuitous 'if'. But it also assumes an 'all', whose unempiricy we have already proved. If it is not the common denominator of all possible words, a common denominator cannot be called necessary, for some world could be possible without it. Besides, logical compatibility also is not an empirical data. It is evident that Kripke forgot that necessity —*that* by means of which he tried to define 'law'— needed to be defined. Kripke wants to define the necessary going through the possible, but he defines the possible as that which is compatible with the law. And there is the law again, which was just the origin of the entire problem. The circularity is manifest, and at the end of the day nothing has been defined.

I do not know why one would want to deceive someone by concealing the fact that when some processes started to be called 'material', this was in contrast with what is 'free'. Men would have not come up with the idea of calling these processes 'necessary' if it were not by contrasting this with his freedom, which is continuously experimented in self-consciousness This was done precisely with the purpose of *denying* that stones, rivers and stars are free. The only meaning of 'necessary' is 'not free'. In fact, we saw here (III, 2) and also in Kant that self-determination is the most intelligible concept that exists. But that is, necessary means not free; this implies not only that freedom is more intelligible that this necessity that scientists look for, but also that this big necessity lacks all intelligible content whatsoever and can only be alluded negatively without us understanding what kind of fixed point this is.

Determinism —as the thesis that all that exists is necessary— is one of those funny statements that in order to have meaning need to be false, for the concept of freedom could only have originated in self-consciousness; i.e., in the direct knowledge we have from a being that is truly free; otherwise, it would be unexplainable the fact that one day the concept of freedom started to be. The thesis according to which everything is necessary lacks meaning if the concept of freedom does not exist; therefore, in order to have meaning, this theses actually needs that some beings are free, that is to say, it requires that the thesis itself to be false. The same happens with the thesis which says that everything is inexistent.

And I am not only referring to the philosophical and thematic determinism of Laplace or Holbach, but also to the 'methodic' determinism which many scientists feel obliged to profess within their disciplines. As we have said, determinism refutes itself in each and every one of its forms. I do not see why physics had to wait for Heisenberg, Bohr and von Neuman in order to bury determinism deep in earth. It was obvious even from before that it was not only a gratuitous and unverifiable thesis, but also that the word necessity itself lacks empirical meaning and hence has nothing to do within Physics.

It is very important to notice, however, that the true concept of *necessity* —not that fictional necessity without content which scientists have pursued in vain—, certainly *has meaning*. Not an empirical one, naturally; we have seen that there is no way to express the necessity of a law in empirical terms. The meaning is something that *must be* and that *has* to be. To be sure, it was obscurely uttered before by frustrated formulations, such as: 'it can not be not being'. But if no empirical content could have given origin to this meaning, the origin had to be self-consciousness. Now, what we know by self-consciousness is freedom. Therefore, the concept of necessity cannot be different from the concept of freedom.

The identity of necessity and freedom is probably the most remarkable feature of the moral imperative, since the moral imperative (III, 7) addresses freedom and, in doing that, constitutes freedom for the very first time. The moral imperative makes the subject free by drawing him to the responsibility of a necessary behavior. As Kant says in precise terms, 'I am free because I *ought*'. Before being responsible I am not free; I am free in virtue of the necessity that imposes to me called imperative: "Neither freedom on its own, as subjective and separated from necessity, is something absolutely real, nor necessity on its own, as isolated, can be called real" (Ästh I 165). "What is real is the unity of (those) contrasts, and we must say that the spirit is free in its necessity; only in necessity it is free; conversely, its necessity consists in its freedom" (EGP 116); "the absolute necessity contains in itself its freedom" (PR II, II 28).

Only due to an absolute misunderstanding of Hegel's thought one can explain the fact that the identity of necessity and freedom has been interpreted as 'known and accepted necessity', granting necessity thereby with a meaning of natural law and determinism. Such necessity does not exist. The necessity of the natural laws "is itself only a sham, false necessity". (PG 118). "But if the law does not have its truth in the Notion, it is a contingency, not a necessity, not, in fact a law." (PG 189)

No one knows what the necessity that has been praised by physics and other scientists —including deterministic psychologists and philosophers of history— means. The only possible meaning of the word necessity is known by introspection, or better said, by self-consciousness, where its content is freedom itself.

"One must not understand by necessity the exterior, but rather the irresistible, the divine, which is an end in and of itself, in relation to freedom" (VG 263).

We saw (III, 9) that this is the moral imperative which, as we saw, is of a divine nature. That is the only possible meaning of the word necessity. As we will see (V, I), this imperative makes man free, for one cannot speak of the autonomy of the self when the course of life is decided by impulses and instincts which were not introduced by the self and appeared miraculously. Only in my positive response to the imperative am I autonomous and free. "The necessity deepens into the concept. And this, which is freedom, is the true of necessity" (PR II, II, 199).

It is shameful to watch the spectacle given by those who distrust of our knowledge and perception of the moral imperative because, they say, it lacks the necessary character that the empirical science and its laws have. Such science *would want* that their knowledge had a necessary character, but they do not even know what it would want, and in fact, such necessity lacks all kind of meaning whatsoever, weather it is empirical or not. Thus, those who disdain the imperative because they prefer necessary knowledge, are going, in fact, after a revelry which does not have meaning nor can it have it. They disregard what is truly known to us and constitutes the only possible meaning of the word 'necessary'.

We mentioned (III 7) the group of 'roles' and 'expectations' that constitute a society, and we showed that its moral character could not be concealed. This system of intersubjectivity —which, as we have explained, is free and necessary— Hegel calls it 'free mechanism' in contrast with the laws of mechanic (the Newtonian mechanics) which in fact lack all definable necessity and are not thus laws. "Only the free mechanism has a law, the proper determination of pure individuality, that means to say, the concept that exists on its own; as distinction in itself, this law is a restless source of movement that lightens itself up and is *free necessity*" (WL II 375).

Here we have the 'origin of movement' Aristotle longed for in his *Physics*. It is the soul that Plato defined as "the movement capable of moving itself" (*Laws* X 896A). What mechanics blindly looked for with its laws was to explain what happens in the world, but it lacked true causality and the production of more being.

CHAPTER IV

Infinite and Distinction

We just showed that the so much praised necessary character of the 'objective universe' as opposed and confronted to the subject, is a deceit because it lacks all kind of meaning. Now, this overwhelming intention and effect has affirmed the infinite character of the physical world, and I believe that when Pascal was frightened because of 'the eternal silence of these infinite spaces', his fear was derived principally from the infinite.

The great public is still shocked by the idea of a space which has been declared for more than three thousand years as infinite. Although the majority of actual physics says that it does not believe any more in the spatial infinity of the cosmos, infinitude is reintroduced in their discourse under the form of the infinitesimal (infinitely small). In addition, must physics still believe in the temporal infinitude of the universe: in other words, they believe that time extends to the past and to the future infinitely, which amounts to keep extolling that the space is infinite for that is the concept of time that physics have. They do this as if the word *infinite* could have any meaning at all in its application the physical.

As it is obvious, we must add that every particularity involves limitation and consequently carries negation to universality. The subject of the infinitude is identical with the subject of universality. No science can do without this subject, let alone logic. It is unsettling and outrageous how one believes that everybody understands what the word 'universal' means. Another disconcerting belief is that it is very easy to give meaning to it by empirical means.

Following Hegel we will show here that, if it is not in terms of self-determination and morality of the spirit, the words 'infinite' and 'universal' lack all kind of meaning.

On the other hand, a true definition of infinite brings a capital problem to philosophy and theology, namely, the distinction between the infinite and the other beings. The illusion that consists in believing that the word "infinite" has some physical meaning besieges theologians too, since they try to distinguish God from the human spirit by means of this term —God and man are identical at least in regard of their spirituality—, as if infinitude consisted in something different than spirituality itself. The definitive clarification that Hegel provided has not been understood by his commentators, despite that it is one of the most revolutionary and irrefutable contributions that a thinker has ever made.

1. Pseudoinfinite

"The infinitely great and infinitely small are therefore pictorial conceptions which, when Lockheed at more closely, turn out to be nebulous shadowy nullities." (WL.I 236).

These expressions are breakouts towards the irrational. One can begin to see this with those who define infinite as 'bigger as anything that we could possible think'. That means, we cannot think; one grants us permission to abandon rationality in the wings of imagination; we are authorized to proceed without concept.

It is time to proclaim to the entire world that the *infinite has nothing to do with a big size.* In general, it does not have to do with magnitude either. Something is infinitely big when it cannot be bigger, but the definition of magnitude is: that which can be bigger or smaller. Therefore, it is a contradiction to talk about an infinite magnitude. If it were infinite, it would be no magnitude.

Likewise, something infinitely small (= infinitesimal) is something that cannot be smaller. But by definitions a magnitude is what can be bigger or smaller. Therefore, an infinitely small magnitude is a contradiction. It is nothing more than fog and shadows. Hegel warns us: "this is a necessary and direct consequence" (WL. I 242).

At this point, mathematicians make a drastic turn and define *infinitesimal* quantities as quantities that can be *always* smaller. But this is mere mambo jumbo. In fact, a quantity is what it is and it cannot be either bigger or smaller. What they mean actually is a quantity that can be substituted or replaced by another smaller quantity. They are imagining that they substitute it for a smaller quantity: Not *always*, (cfr III 9) though; and this is what is decisive. An 'always' is not a sensible data, and consequently, it is not imaginable. Not even the imaginary representation fulfills the definition they are giving of infinitesimal. And the same goes to every definition of theirs that defines big as a quantity that can be always bigger.

When the mathematician or the empiricist says that he imagines himself repeating the operation indefinitely, he is only imagining the beginning of it (five or six times). But the beginning is not infinite. In addition: with the adverb 'indefinitely' the *definiendum* reappears, there is circularity and they have defined nothing. In regard of the concepts, they do not manage to define 'always' nor 'indefinitely'; in regard of imagination, they may 'repeat' but they do not this "indefinitely". Without this last element the definition leaves aside what is essential. According to them, this is what would make the quantity infinite. "This is the bad infinite: when one says, and so it goes *in infinitum*" (GP III 171). "It has been exposed that the indefinite progress belongs completely to the reflection which lacks the concept" (WL II 500); "supposing and determining never fulfill the goal" (JS 27).

What we must reproach to the indefinite process is that it is not *understood*; it only appears to be so; we never get through it to the concept. It is useless to say that Mathematics and Physics understand by infinite an indefinite process, for there is nothing to *understand* there. What we have there is an imaginary construction that is not infinite. It is an abuse of language to call something a thing that it is not. "The infinity of the infinite progress remains burdened with the finite as such is thereby limited and is itself *finite*." (WL I 131).

If it was not an inveterate self-deceit, we would not insist upon it. Perhaps somebody thinks that the concept of infinite could have an empirical or operational meaning in the following way: I could add a stretch to a straight line, and I could add to the result another stretch, and so on.

But this 'so on' is not empirical; the first four or five times perhaps they are, but they do not suffice to build up an infinite; on the contrary, the expression 'and so on' would need to be defined, and its definition would be precisely the definition of quantitative infinite that we were looking for. If they tell us that 'in principle' the operation of adding a stretch is indefinitely repeatable, they are not delivering us any empirical data or observable operation, but only a *statement* whose meaning is not understood so long the adverb 'indefinitely' is not defined, which was actually the only thing at stake here. Furthermore: a given straight line is perhaps an empirical data, but 'any' straight line is not. 'Any straight line' is the same as 'all straight lines' and we have showed (III 9) that 'every' is neither an empirical nor an operational data. Besides, they said that I 'can' add a stretch to the line. The experience that originates the concept of 'can' is not an empirical but a selfconscious one: I can give myself other determinations; I can choose either this or that. The concept of this concept is freedom and self-determination.

It is obvious even for the most superficial reflection that sensibility does not bring us any data at all that could be called infinite and that the origin of the concept of infinite is not empirical. On the other hand, the said concept is evidently not constructed by the negation of empirical finite data, for negation does not add any content whatsoever and, consequently, we would still be missing the characteristic content of infinite, since it was not given by any of the empirical data. Furthermore, the empirical data do not have a label that indicate it is finite or infinite; if they seem to us infinite (and they actually are) it is by contrast to the idea of infinite that preexisted in the mind. Descartes had already explained it. That color is green and period; the green does not tell us anything about finite or infinite things. The sound is sharp and period; the sharp does not say anything about finite or infinite things. We would have never come up with the idea that things are finite if the concept of the spirit did not preexist in the spirit.

"It is sheer irresponsibility no to see the fact that we call finite or limited something contains the demonstration of the *real presence* of the infinite and unlimited, that the awareness of the limits can only exist insofar that the unlimited exists there in consciousness" (EPW 60 A).

The preexistence of the infinite in consciousness is the condition of possibility of all the attempts to grant the word *infinite* some quantitative, empirical, imaginary or operational meaning. The only bizarre thing is that the authors of such attempts do not realize what is it that guides them, for it is obvious that no element of the above mentioned kinds brings up to their minds the idea of infinite; on the contrary, their mind has this initiative has taken their minds and it costs them a lot to invent artificial combinations of such elements that bear some resemblance with the idea of infinite. Besides, the similarity is always false and the attempts fail. Their definitions are either circular or they use in the definer a word or sign which meaning is not as the mentioned ones.

For instance, the rule for the formation of mathematical series (which are the modern version of the indefinite process), includes the letter n, which means 'any', and we already pointed out that 'any' is not an empirical or an operational element.

In addition, one works with 'sets', probably with the belief that a set is an empirical data. Despite the popularity of this trend, it is imperative to realize that empirical things are not constituted as sets nor do they form them. The oblique objects that exist in the world are separated from each other by thousands of objects that are not oblique; our *intellectual* consideration is what delineates the set of oblique objects. We could say exactly the same thing about the organisms called mammals and about the metals called bodies. Therefore, a set as such is not an empirical data. In order to build up a set, it is necessary that it is constituted by all oblique objects, but we have exposed (III 9) that 'all' is neither an empirical nor an imaginary data.

Despite the importance of the contribution he made to mathematics, Georg Cantor —supported by Richard Dedekind— did not say anything new in regard to the fundamental problem we are dealing with. It is true that unlike his predecessors Cantor does not attribute infinitude to the last addend of the series, but rather to the entire series: he attributes infinitude to the sum of all finite elements. No one could deny the sagacity and agility of this radical turn that made Kroenecker fly into rage. But, evidently, no one can speak of a sum or a set without implying the concept 'all', whose meaning is not empirical or imaginary because only the infinite is universal. As a definition of infinite Cantor's great construction is circular: it presupposes that we understand the meaning of the word 'all', which can only be understood in function of the infinite. Of course, he does not ever say a word about the *non-existence* of the series: it does not exist either in reality, imagination, paper or blackboards. Once again, we are dealing with the indefinite process, which we could define as the search of a meaning with the condition of never finding it —in other words, with the condition of never reaching the concept. The reason of not calling the last element of the series infinite is that the last element does not exist. And the same thing goes to the entire series: it does not exist.

Further, all these attempts work with numbers. But a number is not a sensible data. It is an invention of the understanding. Animals do not know how to count; Thomas Aquinas and Kant already said this: sensibility does not perceive the number. Hegel said that too: "Number is a non-sensuous object" (WL I 212).

In order to be convinced of this, we must bear into account that we are able to count the most heterogeneous things, without them having any sensible relation whatsoever between them. A flower, an emotion, a tempest, a flavor, a soul, a bull and a thought sum up seven. Perhaps one could say that the common denominator is that all of them are beings. Now, this would be enough to demonstrate that the number is not an empirical data, for we have exposed (II 7) that empirical data do not apprehend the being. However, we could *add* nothingness as another numeral and then we would have eight. Whenever we have the whim of counting stuff, we do not depend at all on the empirical: numbers have nothing to do with sensibility. The example we just mentioned teaches us that —even when the countable elements are objects that can be perceived empirically— the idea of number does not enter through the senses but is rather an initiative on account of the intellect. Besides, in order to constitute a number, the elements need to be summed up between them; otherwise, each element would exist independently and we would not have the total number. But empirical impressions —as they originally come to us— do not come this way: each of them is what it is and knows nothing about sums. The idea of summing them up must come from someplace else. Not even the fact that five empirical impressions could come simultaneously to our minds would make us count them; let alone the case of successive impressions, for the best thing that mechanical memory could do is to present them at the same time.

In a word, the efforts to give meaning to the word 'infinite' which are not based on self-consciousness employ the number, but for this very reason they fail since the number is not a sensible data. Furthermore, what they build up is neither the infinite nor anything that resembles it. Naturally, they can always employ the arbitrary recourse: we understand *this* by infinite. But we already said that whoever speaks like that remains only with this and renounces thus to the infinite; he prefers this and quits thereby the search for the infinite. Not even he knows why he takes this and not that, for he renounces to look up for the meaning of infinite. He could say I understand 'abracadabra' or 'X' by this, but whoever hears him does not know what he is talking about: he does not even know it himself.

We would now have to proceed positively and disentangle the true meaning of the infinite, but since it is also the meaning of the universal, it would be more convenient to dissipate the common illusion of those who believe that the universal has a meaning which is not intersubjectivity as such.

2. Pseudouniversal

First and foremost, one should notice how inadmissible it would be to build up the universal as the negation of the particular and the individual. First, because negation does not provide any content at all and thus the only positive content would be singularity, which is precisely what we cannot employ in order to construct the universal. Second, because the individual would be defined through the negation of the universal, the process would be circular and nothing would have been defined.

What we have just said refers to the alleged singular that we supposedly perceive by means of sensation. In fact, the empirical data says nothing about individuality or universality, the same way it says nothing with regard to finitude or infinitude. All these considerations are provided by the mind: they are not contributions of the sensibility. On the level of individuality, the fact that the individual must be mentioned in the definition of the universal and vice versa is very illustrative; this demonstrates that, if we leave aside the fixed and unintelligible abstractions of the abstract intellect, the concrete and real individual is universal in itself, and it would be the more universal the more individual it is. But let us not rush too much.

It is important to remind (II, 6) the reader how frivolous the theory of abstraction is. According to this theory, the origin of the universal is empirical data through a mysterious process of generalization. Before making a generalization in order to get a concrete universal, the mind needs to know which are the pertinent data, among the innumerable ones we have in the world, because just from them, not from all the existent ones, could the mind abstract the universal in question. Now, in order to know which objects we need and which aspects of these objects are relevant, the mind needs the universal in question as a *previous* guide; otherwise, it would be trapped by a total despair, it would not know where to begin or where to draw its attention. Far from being the product of empirical impressions and the result of the said process of generalization, the universal is the condition of possibility of the former and precedes the latter.

The aforementioned theory of abstraction is an unforgivable misplacing, because Plato had already showed —in the decisive way we just evoked— that the universal necessarily preexists.

Some people asked: where is it?

The key point is what we have to say. The question is *imagining* the universal (what is imaginary is spatial), *not understanding it*, because only in regard to a spatial object can one say things like location, position, etcetera. And so the same happens as in the case of the infinite: the infinite of imagination is not the infinite, and a universal of imagination is not universal. Neither the infinite nor the universal are imaginable.

By definition, the universal is what is fulfilled in many, in the belonging particulars. But an imagined and spatialized universal falls beyond the singulars; it is distinct and distant from them; it is not in them, it is not fulfilled in them and hence it is not universal; it is in the imagination and the particulars remain outside. The abstract intellect —whose *modus operandi* is to follow imagination quietly— believes that the fulfillment of the universal in the particulars consist in its mere considering that the universals are fulfilled in the particulars. But that would be a fiction of universality, not true universality. Nothing can be legitimately called if it is not effectively fulfilled in the singulars. All the farce would be reduced to say that intellect *has the whim* of calling universal something that is not universal; but a whim does not legitimize the use of the term in the opposite sense contained in its definition.

Of course, it is useless to reply that one can define universal as that which is predicated of many. To predicate is to affirm that something is fulfilled in the thing that is being predicated. This is the only meaning of the operation called *predicating*. All the attempts to define predication by empirical means are doomed to failure. For instance, to say that one predicates something when one has the verb 'to be' is entirely misguided, for all propositions predicates something about something, and the immense majority of propositions are built up without that verb. To sustain that there is predication when the attribute 'goes along' the subject in question is also completely mistaken, because the hyperbaton or other grammatical device can separate them to the opposite extremes of a sentence and yet the predication remains. No empirical data corresponds to the meaning of the predication. Its meaning is to affirm that something is fulfilled in that which is being predicted. And here we are, exactly where we were before this way out came to light.

One should notice that despite the pretensions of the abstract intellect of considering the universal within the particulars, the fact is that this does not happen as tall. What he is considering universal is either an imaginary confusing representation as indeterminate as possible (for all determination would wipe out its alleged universality) or a word that supposedly consists in a set of sounds in a blob of ink. Any of these alleged universals is an entity on its own, which is completely different and distant from the singulars being referred. It is false that the intellect considers or even imagines it fulfilled in the singulars, for it considers or imagines it separately from the objects. It is not even a fictitious universal; it is not universal even for the deceiving intellect that calls it universal.

3. Infinite and universal

Infinite is the spirit. Every spirit, "since there cannot be two classes of reason or two classes of spirit" (PR I 43).

This is a necessary consequence of what we have said: since the concept of infinite cannot be empirical or imaginary, its origin cannot be sensibility or imagination; therefore, its origin is self-consciousness. But what we know by self-consciousness is the spirit. It follows that the meaning of the said concept is the spirit.

When it comes to the spirit, the point of departure has to be —as Hegel affirms emphatically— the immensely meritorious contribution of Spinoza: every determination is negation; every peculiarity is limitation. Nevertheless, Spinoza did not understand that, if the limit is negation, the infinite is negation of negation.

When I published my previous work I had not yet realized that *finitum* and *determinatum* are passive past participles. If we say that someone is infinite, what we are actually doing is denying that this person has been determined, for he is giving himself determinations, that is to say, he is self-determining himself. Infinitude is self-determination.

"The spirit is not natural; it *is* only that which he makes of itself" (GP II 494).

Infinite is what is not determined by nature but what gives itself its own determinations which he can wipe out and replace with other ones. "Freedom consists precisely in the indetermination of the will, that is to say, the will does have in itself any natural determination. This is why it is in itself a universal will" (NH 224). "This absolute unity of individuality and universality called the *self*" (JLMN 163).

"The infinitude is the self-determination which refers to itself, the posing of an immanent and own determination, which is truly something indifferent for he who poses it" (NH 87s).

"The practical spirit (moral) is first and foremost free will in the sense in which the self can abstract from any determination in which he is, and in any determination he remains undetermined and identical with himself" (NH 57).

We showed (III 7) that self-determination consists precisely in moral; the appearances are only natural determinations; self-determination consists precisely in not letting them to determine the self; this task is only for the self to do.

"The awareness of freedom contains the understanding that the subject has of himself as a person, that means to say, as a universal in his individuality, as something which is capable of making abstraction of everything particular and of laying it down as an infinite" (VG 175).

The last thing we said should not be understood as if we were talking as potential infinitude, that is to say, as if infinitude consisted in the spirit's capacity of having all the determinations. It is the other way around: *Determination is limit, negation.* The infinitude is negation of that negation. The spirit is infinite because it does not have determinations "he is only what he makes himself being" (GP II 494).

In itself, the spirit does not have limits, and that is the definition of infinite: that which has no limits. As Hegel says, the limits and the determinations that he gives to himself are indifferent to it. In fact, they do not limit him, because he can do without them whenever he wants. Even in the act of giving itself certain determination or peculiarity, the spirit manifests its infinite and universal nature, for he could give itself this one instead of that one; these determinations do not determine the spirit; it holds the dominion over his own acts. "By virtue of this original unity it follows, in the first place, that the first negative, or the *determination*, is not a limitation for the universal which, on the contrary, *maintains itself therein* and is positively identical with itself." (WL II 241)

It is useless to object that such thing is not what is commonly understood by infinite. By infinite one can only understand that which has no limits. The only sense in which the *definiendum* can be real is the one we have been referring to. We have showed that people commonly imagine something that is *not* infinite, despite the fact that mathematicians and physics have the whim of calling it like that. The concept of infinite that we have showed can only be rejected by someone who has *another* concept of infinite, but neither philosophers nor physics have such concept or can have it. In particular, we said (IV, 1) that one should not confuse the infinite with a big magnitude. Further, one should not be tempted to say that the infinite is a 'magnitude without limits'! By definition, every magnitude is a limit, so to talk about a magnitude without limit equals to speak about a limit without limit. "What has all determinations" is (implicitly) contradictory, because there are determinations that are contradictory between themselves. Besides, if all determination is limitation, this statement would be useless to define what is unlimited and infinite.

Let us do without the resistance of those who are not willing to admit that the human spirit is infinite. Evidently, they think that the human spirit is 'here' or 'there', in a given place and not in the antipodes; therefore, they conclude that it is not infinite. The ship of this objection is so feeble that one big wave will be enough to sink it. On the one hand, they suppose that the spirit is something material and spatial, for that would be the only way they could speak of location and place. On the other hand, they believe that infinitude is spatial; something so big that is everywhere. We have said enough in this regard. Another recourse the objectors could employ is to say that the infinite is that which has all determinations, denying thereby that the human spirit is not in that case. But we pointed out that this could not be the definition of infinite:

> In a word, finitude consists in having a limit, which means to say, that it puts its not-being in that point; in other words, in that point this thing ceases to be and makes reference to some other thing. On the other hand, the *infinite* reflection consists in that I do not refer to anything else but myself and I am an object to myself. This pure reference to myself is the *self,* the root of the infinite Being itself. The leave aside completely, everything

> that is finite. The self as such does not have any content given by nature (immediate), but rather he has only himself as content. This pure form is at the same time his content. Any content given by nature is, in the first place, limited, while the content of the self is unlimited. In the second place, the content of nature is immediate, while the pure self does not have immediate content since it only exists by means of doing without everything else. (NH 221).

"For the characteristic feature of the spiritual is not being abstract, but rather something vital, a universal individual, subjective, which determines and chooses itself" (PR III 47).

> It is true that we sometimes express ourselves like that: my will was determined by such motives, circumstances, stimuli, impulses. At first, this expression implies that I behaved myself passively. But in fact I did not behave myself passively, but essentially in an active way, namely, my will admitted these circumstances as motives; it *made them be regarded* as motives [...] By means of reflection, I can go beyond any determination given by the circumstances [...] Circumstances and motives exert so much influence upon man as he lets them have. (NH 222s).

This self, which is essentially universal, is the one that specifies itself with these particular determinations, retaining its universality thereby since it gets hold of them freely and they do not set barriers to his self because the spirit could easily do without them and obtain other determinations. The really interesting thing about this is that in exercising its freedom the self truly becomes a self, in other words, it becomes more individual. The more universal and free it is, the more individual it becomes; and this comes about in the act in which it particularizes itself; if the spirit would let itself to be determined by natural impulses, it would not be something individual but only another part of nature. It follows that universality, particularity and individuality are identified with what is concrete, subjective and vital, despite the fixational intellect that only deals with abstractions and keeps these three things separated and without understanding them, for the abstract universal is not universal and the abstract individual is not individual. We have seen already (IV, 2) that the abstract intellect does not know what to do with them and does not understand them, for the abstract universal is not universal and the abstract individual is not individual. We saw that this abstract intellect did not know what to do with the universal, which by definition is fulfilled in the individual and identifies itself with it. This mediocre way out consists in declaring that the universal is not real; but that getaway is blocked by the evident fact that if it is not real it is not universal, because we cannot say that something unreal is *realized* in the individual.

There is nothing weird about the fact that the spirit is the meaning of the universal. If no empirical or imaginary data is universal, it follows that neither sensibility nor imagination could be the origin of the concept of 'universal', and hence the origin of it lies in self-consciousness. But what we know in self-consciousness is the spirit! That is the content of such concept.

By the way, it follows also that the universal and the particular identify themselves, for when the spirit knows itself by means of self-consciousness it knows himself as a self, that is, as individual. This clarifies the explosive Hegelian thesis according to which judgments can only be true in regard to the spirit; the only thing that the judgment affirms is that the individual is universal:

> It must be considered a complete lack of observation the fact that one does not points out in logics that *any* judgment emitted contains this assessment: "the individual is the universal" [...] It is true that the determinations of individuality and universality, of subject and predicate, are distinguished, but this does not deny the general *fact* that any judgment declares them *identical* (EPW, 166A).

The contemporary logicist efforts miss the target completely when they say that the judgment only affirms that the subject is an element of a given set. We pointed out (IV, 1) that a set is not an empirical data and that it does not exist in reality. The only thing the spirit has is concepts, and only in the level of concepts one can give a meaning to *every* judgment. One has to do without the peculiar and the incidental character that the subject or the predicate of the judgment might have. Now, the common meaning to all these judgments is the one that Hegel stresses: the individual is the universal. That is the true form of the judgment: Hegel is right when he says that common logic is not formal enough in spite of its pretensions of being formal.

But let us go back to the infinite, which is our central subject. We have seen that the meaning of infinite is self-consciousness (the self), which is first and foremost self-determination since it is not determined by nature. But we said (III, 7) that self-consciousness and the consciousness I have of others feed each other, they make the other exist, and one can truly say that one is the being of the other: "without a thou the self is impossible"(JS 378). The most elemental analysis shows us that the meaning of infinite is self-consciousness, and if self-consciousness is essentially intersubjectivity, *the very concept of infinite includes distinction*, that is to say, it is not a monistic and/or a pantheistic reality, as rightist and leftist superficial critics naively affirmed.

On this point, one cannot be too radical. As we have seen, one cannot give any other meaning to the word 'distinction' or 'distinct' than that which consists in intersubjectivity. It follows from this that not only the imputations of monism lack *meaning* completely, but also that this contribution of Hegel is one of the most revolutionary and important ones ever made in the history of philosophy and in science in general (including physics, of course).

4. Imputations

If the leftist counterfeiters would have wanted to respect the rigor that science imposes, they would have put more attention to texts like the following one: "The Enlightment came from France to Germany, but with the difference that nobody dealt so openly with the dogma of such knowledge; they were twisted and turned in order to maintain the apparent respect towards religion, something that is still done now" (WG 916s).

It is clear in this text how genuine is the contempt that Hegel has against the hypocrisies that cover up and dissimulate the true extent and meaning of a philosophical doctrine with the purpose of not offending certain religious instances. Besides, it is a text of Hegel's last intellectual period; it is absurd to suppose that they could impute him with the same charge he holds against them. I do not see how Findlay could possibly hold the following posture: "Though Hegel has veiled his treatment of Religion in much orthodox-sounding language, its outcome is quite clear. Theism in all its form is an imaginative distortion of final truth" (1958, 151). What we have here is this: the prestige and the huge intelligence of Hegel are so imposing that the atheists do not give up their wishes of having him on their side. One can read similar forced co-options in Kaufmann, Lukács, Althusser, Marx, Feuerbach, Adorno,

Kojève, Marcuse, Garaudy, etcetera. One does not need to make a big fuzz about this: one only has to distinguish between the publicist and the scientific genre, for the form really has to be subjected to objectivity and to the accuracy of the texts.

Since we mentioned Enlightment, it is worthwhile to read this other passage from Hegel: "Enlightment —that conceit of the abstract intellect— is the fiercest enemy of Philosophy; it hinders that one makes reason patent in the Christian religion; it hinders that Philosophy demonstrates that the testimony of the spirit of truth is placed within religion" (PR III 225).

I do not believe that a careful exegesis could question the sincerity of this declaration of war against Enlightment and in favor of Christianity. If Findlay *et al* would have said that theism is false and that Hegel was dumb —or, in either case, that he did not see the consequences of his principles— we would be facing a whole different problem, but to say that Hegel is only pretending is something that cannot be done if one examine texts so aggressive and militant as the last one we quoted.

This other passage says clarifying things in regard of the illuminist pandemic:

> That was in the rationalistic, enlightened, and apparently rational religion, but it was nothing else than renouncing to the fundamental teachings of religion, so that everything remained in general and superficial doctrines, such as that there is a God, that a man appeared, Christ, in order to teach de divine commandments and in that sense he appeared as a divine man, etcetera [...]. A religion that is satisfied with such contents is as insipid as the belonging philosophy [...]. An agreement in which everything is so squat and flattened does not satisfy the depth of religiosity or the depth of the reason that thinks (EGP 292s).

This passage does not need to be commented. Findlay and his followers lacked the scientific seriousness of the hermeneutic: that is all.

There is a point which is closely linked to the subject of our present chapter in which the aforementioned gentlemen would like Hegel to change his mind: Findlay expresses it thus: "Hegel further says that, in his True Infinite, the element of Infinity will absorb and overcome the Finite, but he might equally well have put it the other way round." (1958, 164)

We will later see that the divine does not absorb the human; on the contrary, the former affirms the latter and makes it exist. But Findlay and his henchmen would want Hegel to think that man was the only thing real, and that all the non-sense about the infinite and the divinity were absorbed by the human. They probably missed the Hegelian mockery about the Greek religion in their reading of the *Phenomenology*: "both gods and men did one and the same thing. The earnestness of those divine powers is a ridiculous superfluity, since they are in fact the power or strength of the individuality performing the action." (PG 508). In contrast to the God of Christianity, Hegel mocks the Greek divinities for not having personality and real individuality: "They are, in true, endowed with the form of individuality, but this is only in imagination and does not really and truly belong to them; the actual self does not have such an abstract moment for its substance and content" (PG 517s). Let us keep into account that these two texts are in the *Phenomenology*, which insistently isolated unscientifically from the rest of the Hegelian work, by those who forcefully want to suck the divine into the human.

Hegel denounces that the Indian divinities also lack real subjectivity and autonomy in front of the human subject:

> In his essential determination Brahma remains the abstract being, the universal, the *substance without subjectivity* in itself; therefore, it is not *concrete*, it is not the spirit (just as it happens with the modern philosophers, who determine God as concrete when they call him the *essence* of the essences). With such content —which is in fact lack of content— that *masculinum* (Brahma) is not an individual subject; the personality is in him an empty form, a personification. It is of utmost importance while studying religions to distinguish between the *mere personification* of a god and a god, something that one can find in every mythology, and the personality, which is by content. Since personification is something superficial, the objective autonomy of the god before the subject collapses. For instance, at the beginning of the Iliad, when Eros or Pallas prevents that Achilles draws his sword, we immediately take that as the subjective feeling of love, as the good sense that makes itself present in Achilles himself (BS 186).

In its parenthesis, this passage denounces those Western men who are incapable of conceiving God as a true subject. What Hegel rejects is a divinity that lacks real autonomy in front of the human subject. What Hegel explicitly and sharply denies is the absorption that Findlay and his followers want to attribute to him.

Furthermore, Hegel accuses of atheism those Western theologians that do not conceive God as spirit.

They accuse Spinoza of atheism when in fact the intention of Spinoza is not to deny the existence of God but the existence of the world (acosmism). But this is true: insofar Spinoza dies not conceive God as spirit, he is an atheist in that sense, but many theologians are atheists in that sense too. "Spinozism if far from being atheism in the ordinary sense, but in so far it does not conceive God as spirit, it actually is. Now, in this sense, many atheists are theologians that only call God the omnipotent and supreme being […]" (GP III, 195), or that reduce Him, as he says in the *Encyclopedia*, to the "undetermined suprasensible" (EPW 73).

When one sees that Hegel even calls the theologians 'athetists', one realizes that the attempts of making him an atheist are only literary attempts which lack seriousness and scientific method. Hegel adds himself expressly to the aphorism "a bad philosophy draws one apart from God […] and a true philosophy draws one near Him." (Rph XXIII) He explicitly warns us that "men did not have to wait for philosophy in order to receive the truth and awareness of it." (PR I 299) While making history of the 3th century of our era, he approves the dictum of Tertullian: "nowadays kids know of God what only the greatest sieges of antiquity knew" (GP II 498).

Against our principle of examining the matters in itself and not the interpretative questions in regard of Hegel's mind, it has been necessary to make a brief pause to analyze the latter ones because, if the reader thinks that the author is pretending, all the reference points start shuddering and turn out precarious, and the intellection becomes impossible. The idiotic accusations of atheism and pantheism have prevented the world from receiving the most profound and true philosophical message that exists. We will demonstrate now that the accusation of pantheism is false, first formally and then in its content. After we have settled some issues, it will be better to leave aside the interpretative pursuit and go to the matter itself.

With the same force with which he criticized the lack of consistency and true subjectivity of the mythological gods, Hegel criticizes the lack of consistency and true subjectivity of the human spirit in the non-Christian religions:

"The Parsi place the bodies of the dead exposed to the open air so that the birds would eat them; in so far as the soul goes, they thought that it thins into the universal" (WG 496). "The Indians have also a very gloomy conception, since the last stage is for them the transition to the universal substance" (WG 495). Even in Judaism "the subject does never reach to the consciousness of his autonomy; that is why we do not find among the Jews the faith in the immortality of the soul, since the subject is not existing in and by himself" (WG 457).

In contrast to this, "the Western infinitude and joy of the individual is conformed in such a way that the subject remains in the substance, that it is not debased, that he does not see himself as a slave that depends on the substance, destined to annihilation" (EGP 232).

"While in the oriental consciousness, the most important thing lies in the fact that the universal is the truly independent, for us in the Western consciousness the individuality of things and of men is beyond everything" (PR II, I, 128).

> In the oriental religions the fundamental situation is that only *the true substance as such* is what is true and the individual does not have any value at all in himself; he cannot get if while he remains in front of that which is in and by its own; the individual can only have value if he identifies himself with that substance in which he ceases to be an object and fades away into unconsciousness (GP I 140).

To tell the truth, these formal statements of immortality and of true subjectivity point already to the content that decides everything: the only thing on which the accusers of pantheism should focus is if individual self-consciousness is preserved or not. Every other lucubration ends up confusing spirit with matter:

"The Greeks did not seriously take into account what we call immortality" (Ästh II 572).

On the contrary, according to the Christian conception, "it is the individual, the real subject, in its intern vitality, what has infinite value" (Ästh II 568).

"Plato did not know how to acknowledge or to conciliate with his ideas the willing, the wanting and choosing of the individual" (GP II 129).

"One cannot say that Greeks understood death in its essential meaning (Ästh II 571).

According to Hegel, the conception of the Easterners incarnated in the system of Spinoza reappears in the Western world: "It is the oriental conception which is formulated by Spinoza in the Western world for the first time" (GP III 165). And Hegel says the following of that philosophical system:

> What one can reproach to that philosophy is that it conceives God only as substance, not as spirit, not as concrete. The autonomy of the human soul is thereby denied; in the Christian religion every individual is conceived as destined to joy. Here, on the other hand, the individual of the spirit is only a *modus*, an accident, something which is not substantial (GP III 196).

One cannot ask for more explicit expressions against any absorption of the human in the divine. And they are in the core itself of the Hegelian system, namely, in the difference between the concept of substance and the concept of spirit. Hegel would have wondered himself about the language one should employ in order to be understood.

5. Distinction

Let us go to the problem itself. Something really simple has happened: those who accuse Hegel of pantheism believe that they can establish a distinction between God and the creature that Hegel cannot establish, but neither they nor Physics —nor common-sensed people, for that matter— have realized that the only possible meaning of the term distinction is the one presented and defended by Hegel. And the same goes for the words *identity* and *individuality*. Fortunately enough —materially constrained by the experiments themselves— quantum physics has started to question if these expressions have any physical meaning at all. Scientists would have spared themselves an entire century had they read Hegel more carefully.

"According to the concept, the distinction does not have any physical meaning at all" (GP I 206).

"In the sensible things there is no true objective distinction, only in the spiritual" (GP I 315).

"What is not distinguished in thought is not distinguished" (GP III 246).

Everything becomes clearer once we realize that it is impossible to give an empirical meaning to the word *distinction* or *distinct*. It follows from this that one obtains the meaning of this concept through self-consciousness. But self-consciousness is intersubjectivity (cfr III, 7), distinction among people. The original meaning must consist in "vital and spiritual relations" (WL I 335).

The most frequent (and funny) thing is to believe that the *distinction* between two beings consists in that one is 'here' and the other is 'there', believing one can pinpoint them with his own finger. It is ridiculous that whoever talks this way believes he has given an empirical meaning to the word *distinction*.

But so it turns out that Gorbachev and Reagan are not distinct, since they are not in the places pointed out by the finger.

These empirical tricksters would answer: I am not referring exactly to the two places I am pointing at, but to two different places, regardless of which ones they are.

The reader perceives immediately that what we are given is not an empirical sign but a phrase, and that our definer has failed in his attempt of indicating an empirical data as the meaning of the word *distinction*. And something even worse has happened: he is saying that being consists in being at two different places. The *definiendum* reappears in the word *different*, which was what one tried to define in the first place. This dictum pretends to be a manual book that tells us how to find out if there is distinction or not, but in order to be effective the manual book itself requires to be previously understood, and for that to happen the word "distinct" has to have a meaning first, but the whole enterprise had as its very purpose finding that meaning! Consequently, we are just like in the beginning, and the allegedly empiric maneuver has proved to be barren.

What such a definer is saying is that the distinction is a visual data, for he supposes that one only needs to opens his eyes to verify it. But if that were so, one could not distinguish two sounds, two smells, two flavors, etcetera. Likewise, the electrons could not be distinct from each other, since they are not visible; and that certainly goes against the intentions of our definer. Needless to say: theologians could not use that distinction in order to say that God and the human spirit are distinct.

Even if we do without the visibility, it is well known in Physics that two electromagnetic fields could be present at the same point of space. According to our definer, these two electromagnetic fields could not be distinct. But in the present context the visibility is decisive, and it is important to notice that the *visible* limits of a body —its empirical distinction in respect to other bodies, so to speak— are not reliable data in the natural sciences at all, since the constitutive fields of this body go far beyond the allegedly visible limits of it and even penetrate the zone of space occupied by other bodies. Therefore, the alleged visible distinction cannot be the real distinction by any means.

But those who try to give an empirical meaning to the word *distinct* by means of 'here' and 'there' incur even in a bigger mistake. The 'here' is so metaphysical and unempirical as Newton's absolute space and, in fact, as any other space. When someone thinks that he is *seeing* the space that separates him from the wall before him, what he is seeing is the wall before him, not the space. The space is invisible, and a point within space is even more invisible, for that matter. The 'here' is a point or a region in space; but none of these things are seen. And if someone believes that he can touch the space with his hands, the only thing one can tell him is that he is not touching anything. If our definer pretended to attribute an *empirical* meaning to the distinction by a 'here' and a 'there', his failure could not be more disastrous.

Besides, if our definer reflects a little, he has to recognize that the idea 'two beings really distinguish themselves' is not the same as the idea 'this being is here and the other being is there'. As Hegel says: "In the here and now as such does not consist the distinction" (GP I 315). Let us make the following question to clarify this: if two beings identify themselves with each other, would they cease to be two bodies? The fact itself that we *understand* this question implies that being distinct is not the same as being in different places, for we understand the difference between two bodies occupying the same space and two bodies identifying themselves with each other. Even he, who is inclined to respond affirmatively to the question, has to understand it first: this implies that the meanings of 'identifying' and 'being one body' are not the same. Now, this will do: the *meaning* of distinction supposedly consisted in being at different places, and we have proved that this is not the case.

The same question we made with regard to identity, one should ask it in regard of a 'twofold location'. The question would be: if a body were at two places simultaneously, would it cease to be the same body? The simple fact that the question is intelligible implies that 'being at two different places' does not mean 'being two and not one'. I owe anthropologist Leslie White the following information: "...it may be remarked that normal children and many primitive peoples find nothing wrong with the notion that a body can be in two different places at the same time" (1964, 279 n.). This fact will do to demonstrate that the *notion itself* of being at two different places *does not mean two* different bodies.

In addition, one should know that the true extent of the experiment of Young and Taylor —to which we will refer later on— is that terms like distinction, individuality and identity do not have any meaning in Physics. An example of this is that an electron can be in two places at the same time. Many physics would want to reject this. They reject exactly what they understand. It follows that the concept of two and not one, that is to say, the concept of distinction, does not mean to be in two places at the same time. Let's repeat with Hegel: "In the here and now as such, it does not consist distinction" (GP I 315)

6. Individuality

Now, if the concept of distinction does not have an empirical meaning, the concept of individuality and identity do not have it too, for it is obvious that individuality implies that a being is *distinct* from other beings, and identity implies that a being is not *distinct* from itself. By some way or another, the concept of *distinction* enters into the definitions of individuality and identity, terms whose meaning is not empirical. And vice versa: the distinction between two beings consists in that they are not identical. If some of these concepts are unempirical, then all of them are. This is precisely relevant in regard to the attempts of defining individuality through the localization in space: the individuality of a being does not mean that this being is 'here' and not 'there'.

The physic Bernard d'Espagnat has drawn some systematic observations from Jean Piaget about the epistemological development of babies —conclusions that are undeniable in this sense. To identify the identity and the individuality of an object with certain localization in space is a practical construction that is useful to the kid in order to coordinate his movements and to integrate his early 'vision' of the world'. It is an implicit *assumption* that the baby makes to orient himself, but this does not have a greater probability than the contrary assumption. "This shows", according to d'Espagnat, "that the idea according to which any macroscopic object necessarily occupies some definite region of space —to the exclusion of other regions— is not an obvious (and hence unquestionable) truth, but rather an element of the definition, useful in given circumstances, of the word 'object'." (1976, xx)

Alongside the psychological developments of Piaget, one could recourse to the undeterministc physics of Heisenberg, Bohr and Von Neumann, which fortunately does not find today serious opponents among physics. D'Espagnat affirms:

> ...the fact that the notion of localized objects is but a construct liberates us from the so-called commonsense view [...] according to which individual macroscopic objects obviously —and therefore *necessarily*— exist as individuals independently of ourselves; hence it liberates us from the apparent necessity of considering them as more basic than numbers, logical structures, and so on. (*ibid.*, xxi)

We have to point out "as individuals" we are not dealing with the existence of material.

Physics could have spared themselves the detour that goes through Heisenberg and Schrödinger (and Piaget) if they had read the section about the sense-certainty at the beginning of the *Phenomenology*. The 'here' and the 'now' only manage to denote something individual if we understand them in function of the subject that pronounces this words. They express thus

> "our intention" (*unsere Meinung*), as commentators like Lauer, Mure, Bonsiepen and Heinrich have very well understood. Therefore, to attribute an individuality to the material objects that does not depend on the knowing subject, not only means to draw a gratuitous and empirically unjustifiable stamen like that of absolute space, but is in fact a statement that lacks all sense whatsoever. Hegel says "there is no distinction between the atoms" (GP I 362).

"Spirit is, in a much deeper sense, this *one thing*" (WL II 121).

Two things are important: the concept of *identity* and *individuality* lack empirical meaning, and to predicate both things of material things is a thesis which has no meaning.

Hume already remarked the first point: First, "As to the principle of individuation; we may observe, that the view of any one object is not sufficient to convey the idea of identity [...] On the other hand, a multiplicity of objects can never convey this idea, however resembling they may be supposed" (Treatise I, IV, ii). The last thing is obvious; the similarity is not only identity but negation of identity; in order to be alike, the objects need to be two and not one. But the vision of a sole object does not suffice either to evoke in the mind the idea of identity: this is a very abstract contribution on account of the intellect. If we explain to a farm worker what we want to say, he would agree with us in that the rock we have before is identical to itself and is individual, but he would have not come up with such a round-about idea despite he has been seeing that rock for years. In fact, despite how much time we spend observing an object, none of its colors, forms or visible details denote identity or individuality: "the one cannot be seen, it is an *abstractum* of thought" (GP I 358).

"The word *this* expresses precisely that distinguishing and singling out something is a *subjective* designation, which lacks something outside from itself" (WL I 104).

It is frequent and coarse to believe that the individuality of a macroscopic object consists in its having visible and palpable limits that distinguish it from other objects. But we said that the existence of fields and forces that constitute each body go far beyond such empirical limits. This refutes by itself such a belief. In regard to the identity of a body across the course of time, it is obvious that senses only witness at best that the body is exactly the same. That it is in fact the 'same' than a minute ago. This is something that empirical impressions do not tell us. Selfsameness is an idea tremendously metaphysical idea, a refined contribution of which the senses know nothing; an (ungrounded) projection towards the material of the self-sameness of the knowing subject, in which the word *selfsameness* does have sense.

If the observed body would have ceased to exist, and if in its place an exactly same body started to exist, the testimony of the senses would not vary at all. This demonstrates that the senses do not know anything about identity and individuality.

Besides, what difference would there be between a body that remained the same and a body that was replaced by other? The selfsameness of the material does not have any meaning at all. I am not only saying that our empirical perception would not notice anything; I am saying that it is completely indifferent that the body has been substituted for another, and consequently, I affirm that we are extrapolating a concept to the physical that has only meaning in self-consciousness, intersubjectivity and morals. That matter is a principle of individuation is one of the most foolish things that have ever been said in the history of thought. It was originated in the illusion of the localization that was dispelled above. Karl Rahner, the most intelligent Scholastic of our century, rejected this impossible doctrine:

"Moreover, identity is given to us, now and in the future, by the identity of the spiritual subject of freedom called soul" (Schriften XII 461s).

The macroscopic is apparent. Its origin is the subjectivist peculiarity of our senses. If the material possessed identity and individuality, this would have to be in the microscopic level. But the most intelligent physics have noticed that "there is no distinction between the atoms" (GP I 362) just as Hegel thought. Besides d'Espagnat, it would be useful to quote other physics, for instance, P. W. Bridgman: "The elemental processes or 'objects' do not have individuality or identificability, nor can they be repeated. The concept of 'sameness' does not apply in the microscopic domain of quantum phenomena (Schilpp II 1970, 346).

Likewise, Eddington says:

"...distinction of individuality, if it has any meaning at all, has no bearing on physical manifestations" (1978, 175).

The next paragraph of Max Jammer summarizes the thought and the experiments of Heisenberg, Hund, Denisson, Wigner, Heitler and London

> These results not only lent weight to the concept of like particles; they also showed that like particles may be indistinguishable, that is, may lose their identity, a conclusion which follows from the uncertainty relations or, more precisely, from the impossibility of keeping track of the individual particles in the case of interactions of like particles. For, contrary to classical dynamics, trajectories could no longer be defined as sharp nonintersecting world-lines but had to be conceived as *overlapping* each other. In fact, all papers on exchange phenomena and, in particular, the calculations concerning the ground state of the helium atom, in which the wave functions of the two electrons overlap completely, showed clearly that the classical principle of an unrestricted identifiability of particles had to be abandoned. Moreover, it was possible to show that Heisenberg's uncertainty principle was already contradicted by the idea of an approximately continuous sequence of atomic-configuration measurements, designed to identify electrons in lower-energy states and hence requiring positional uncertainties smaller than average electron distances. (1966. 344)

This is related with the 'forbidden distance' or impossible orbits of Bohr's atoms which we will in short bring into consideration. Let us only look first at this quotation of Paul Dirac: "If a system in atomic physics contains a number of particles of the same kind, e.g. a number of electrons, the particles are absolutely indistinguishable one from another. No observable change is made when two of them are interchanged." (1981, 207)

Hegel had already said in times of classical mechanics that we are dealing with "exterior objects, not individual ones" (WL II 376).

In the atomic model of Bohr, the atoms 'jump' from one orbit of radio to another depending on how much energy each of them has. However, as it was experimentally demonstrated by Planck's discovery of the indivisibility ad infinitum of physical realities, an orbit's radio cannot have any dimension, there are radios that are physically impossible, which means that it is impossible for the electron to be at such a distance from the nucleus. Therefore, it is not the case that the electron *passes* from one orbit to another, but rather that the intermediate states cannot exist. The only thing that happens is that the electron disappears from one orbit and reappears in another. But how could one affirm that it is the 'same' electron, if there is no continuity between one state and the other? Further: is it relevant at all that it is the same electron or not? The only possible meaning of being the same is considering oneself to be the same. And this happens without any substratum. It is not the case that the subject remains the same even when he is not conscious of being the same; we already showed (III, 6) that these intermediate states do not exist. I am myself *because* I consider myself to be so. This lies in mutual dependency with intersubjectivity (cfr. III, 7): Just as it is true that I consider myself the same in function that the others consider me the same, so it is true that they consider me the same because I consider myself to be the same. "Only in morals this concept of the absolute individuality of consciousness has properly sense." (GP I, 271). Individuality and identity of matter, is an absolutely unjustified projection of concepts which only have a meaning in the spirit.

In radioactivity, the tunnel effect demonstrates also that it is absurd to talk about the identity of the particles. There is a 'prohibited zone' for the particles around the nucleus, but these particles are in fact emitted from the nucleus towards the surrounding world. The zone is prohibited because the kinetic energy of the particle would be negative in it and its speed would be imaginary. Therefore, to say that a particle passes by this zone is to utter something impossible. All the problematic that may rise up in regard of the identity of the particle emitted by the nucleus is based on the ignorance of what the physical world is in contrast with the spirit.

This lack of knowledge was what made Einstein broke up with Bohr, and it is also the reason why he was incapable of accepting the decisive *facts* of relativity and, more in particular, the disturbing fact discovered by Young's dispositive and Geoffrey Ingram Taylor's technique. This technique is described by Ted Bastin:

> ...any interference experiment can be conducted at such low intensity that effects due to statistical assemblages of particles can be discounted, yet interference occurs just the same. Hence even in the traditionally central case of electron interference, purely statistical treatment is not possible. (1971, 5 n.)

In other words, it is not the case that a photon is interfered by another, but rather that it interferes with itself. And the same goes for the electrons. Cfr also: EB 23, 20, 2.

Niels Bohr summarizes the problem thus: ""...to be obliged to say, on the one hand, that the photon always chooses *one* of the two ways and, on the other, that it behaves as if it had passed *both* ways" (1958, 51).

This is the dispositive: to the left hand we have a fountain of photons, in the center we have a partition with two orifices or little slots, and to the right hand we have a screen where the light is reflected and where we can observe by the form of the lights and the shades if there is interference or not, for strips of maximal intensity juxtaposed to dark strips mean that two wave trains are being superimposed, whether in phase or not. Let us remember that by the low intensity of the fountain one has achieved monophotonic rays; in other terms, it is not a front of various photons that goes toward the partition, but a single strip of a photons in which each of the elements advances one at a time. Now, if there is individuality in the material, the photon can only go through one of the orifices; but then there would be nothing of interference, since there would not be two different trajectories and hence there would not be superimposition or annulment of wave crests, something for which two wave trains are required.

The phenomenon which has caused an unmendable consternation among physics is that the interference is produced when the two orifices are open, but if we block one of them, no interference is produced. The intellectual honesty of Bohr and his colleagues oblige them to say: the photon, evidently, goes through one of the orifices, but it behaves *as if* it went through both of them. Otherwise, they would have to say something much more scandalous and completely absurd: when passing through one of the orifices, the photon *knows* if the other orifice is closed or open, and as a result, it creates or not interference on the screen.

It is obvious, however, that an *as if* does not solve the problem and hence the Physics of our century has not dared to face the true logical conclusion of the experiment. The conclusion is that individuality, distinction and identity are categories that have nothing to do with the empirical word. If the entire physical world was only *one being*, the observable phenomena would not be different as how they really are.

The experiment has been repeated thousands of times and in the most diverse circumstances and variants, always with the same shocking effect on the minds of the classical physics. For instance, if we capture one of the two resulting photons of the decay of positronium by means of a polaroid in a vertical position, the other will certainly go through a polaroid in a horizontal position. But if we turn forty-five degrees both polaroids, classical optics has to affirm that the first photon has fifty percent of probabilities of going through and that the second photon, which knows nothing about the former one, would have the same probabilities. However, things are not this way. Let us put two polaroids in a straight angle: if a photon goes through one of them, the other photon will always go through its corresponding polaroid. It acts is as if it knew that the other photon has gone through. If we do not want to sustain the absurdity that a photon, once it has went through its polaroid, communicates with the other and orders it to go through its corresponding polaroid —a communication that would have to take place at a speed faster than light—, the only solution is to say what physic O.R. Fisch has affirmed: "We must consider the two photons as being one system" (Bastin, 1971, 20).

It is obvious that we should repeat what we have said about the *as if* of Bohr. To begin with, physics should have never considered the two photons as two separate beings. They ought not to be deceived by the belief that individuality has some meaning in physics. Einstein accused quantum physics of believing in telepathy. But every problem of communication or telepathy supposes gratuitously that they are 'distinct' and 'individual' entities. Physics are ridiculously assuming this pseudo-problem, because these concepts only have a meaning in the spirit, not in matter. The 'occult parameters' of Einstein, Podolsky and Rosen are only efforts to solve the same pseudo-problem.

7. THE BOTTOM OF THE PROBLEM

After demonstrating that it is impossible to give empirical or physical meaning to the words *distinction*, *identity* and *individuality*, we are now in the conditions of going deeper into the matter.

We said that those who accuse Hegel of pantheism should only focus on whether self-consciousness is preserved or not. And the bottom-line is this: this ideal and not physical contraposition between different 'selves', between self-consciousnesses that reciprocally address each other and make each other being, is the only possible meaning of the word *distinction*. Evidently, however, the ideal and not physical entity they create thereby is only one; there is no way to fragment it because it is not material. There we have distinction in the identity and vice versa; it is impossible to define one without the other in the definition. The great difference between the abstract intellect that does not understand and the reason that understands consists in that the former "commits the same absurdity of making that which is pure relation into something devoid of all relation." (WL. I 211).

"The elements of the distinction are the identity and the distinction itself" (WL II 34).

It has been nothing more than sheer superficiality to attribute unintelligibility to the Hegelian thesis according to which "the identity of the identity with the no-identity" (JS 96), for "the identity [...] does not exclude the distinction; it has it essentially in its own determination" (BS 371). The ideal entity called intersubjectivity cannot exist if these poles are not distinguished.

The abstract intellect stiffens: it considers that identity and distinction are things that have nothing to do with each other and are not related at all. One *misunderstands* thus both of them, since it is impossible to give them meaning by empirical or physical means. The abstract intellect uses these two words without giving them meaning.

In contrast with reason, " in general it stands for the understanding as abstracting, and hence as separating and remaining fixed in its separations. Directed against reason, it behaves as ordinary common sense and imposes its view that truth rests on sensuous reality, that thoughts are *only* thoughts, meaning that it is sense perception which first gives them filling and reality" (WL.I 26)

We have long enough demonstrated that this opinion is something utterly impossible. The sensible perception cannot give meaning to the concepts. What happens is that the abstract intellect 'thinks in a similar way than imagination', as Hegel denounces. For imagination, each thing stands on its side with no relation to others; things are unrelated within the (imaginary) space. Properly speaking,

> thought, mediation, the intellectual determinations as such are forms of relation, but they can be immobilized as imaginary figurations, and just then we must think properly in the form of imagination: the intellect (PR I 293).

But:

> this holding apart of likeness and unlikeness is their destruction. For both are determinations of difference; they are relations to one another, the one being what the other is not; like is not unlike and unlike is not like; and both essentially have this relation and have no meaning apart from it (WL. II 36)

The ideal and mental reality called intersubjectivity —which is more real than anything physical, as our third chapter showed— is a simple entity because it is immaterial; it is identical because *it consists* in that its poles are distinct from each other; it *consists* in that every self-consciousness addresses the others directly and reciprocally. This reality is pure act, self-determination of each of the subjects in action and a movement that integrates every one of them; but the intellect separates and makes these things static. It "commits the same absurdity of making that which is pure relation into something devoid of all relation." (WL I 211)

"If we take into account [...] that the simple does not need to be void, one dispels the aforementioned impression, as if the simple excluded from itself the distinct" (EGP 277).

The whole point is not to confuse simplicity with nothingness. Intersubjectivity is simple, because it is not material; but it consists in the distinction that exists among the subjects, which is the only real distinction that there can be. According to Hegel, the identity in Aristotle is "activity, movement, repulsion, and therefore it is not inert identity; identity is identical to itself in the distinction" (GP II 164). If the identity is not nothingness, that means to say, if it has content, it contains *ipso facto* a distinction.

By mere conceptual analysis, Proclus affirmed rightly that the multiple (i.e. distinction) is not in itself; that one can only understand the multiple as included in and streaming from the one (i.e. identity); the multiple would not be multiple if they were not one. Indeed, if they were completely unrelated and had nothing to do with each other, we would not call them many, but we would call each of them one by their own, and the mind would not know of the other when considering each of them. The multiple must stream from the one itself; the distinction has to be included in the identity. For that reason Hegel says: "The reality of the contraries and real contraposition can only be brought about by identity" (JS 97). This could only be rejected by whoever manages to give the words *'identity'* and *'distinction'* another *intelligible* meaning; not an imaginary one, for we saw that what one can imagine is empirical and no empirical data can correspond to the meaning of these words.

For that reason, Hegel says the following about those who accuse him of pantheism:

> These doctrines are of speculative kind, and if there are theologians who are not able to understand them, that in this point they cannot follow the pace of the concept, they should at least, in their weakness, leave alone these speculative subjects. Theology is to understand the religious content; these theologians should acknowledge that they cannot understand, and they should refrain from passing the judgment of pantheism to the concepts they have not understood (PR I 256).

The first thing that these theologians do not understand is the word *God*. As we saw in our previous chapters (III, 1), they do not know what the spirit is and they tend to define it as the negation of the material. This recourse is pointless since the *definiendum* does not get thereby any other content than materiality itself. And we saw (IV, 3) that they do not know the meaning of 'infinite' either. This should not surprise us, since the meaning of infinite is the spirit itself. "We, on the other hand, consider God as spirit" (PR I 8); now, "the spirit is essentially being for the spirit" (PR I 201). The spirit is intersubjectivity; but, as we just exposed, there is the identity, and not only does it not loose distinction, but rather there is precisely the identity so that the distinction between God and men can have any meaning.

"This relation with the Absolute is not suppressing both, for there would be no distinct thus, but rather they must remain distinct without losing that nature" (JS 96).

"The principal is the true determination of this identity, and the true identity is the one that exists in the infinite subjectivity, which cannot be conceived as neutralization or reciprocal cancellation, but only as infinite subjectivity" (PR II, II 68).

> In universal history, the only thing that matters is that these two parts remain in absolute unity, true conciliation, a conciliation in which the free subject does not succumb in the object mood of the spirit, but rather obtains his full autonomy, and in the like way, the absolute spirit, the objective and reluctant unity, obtains its absolute right as well (VG 244).

This subject about identity and distinction is the point in which the famous *'conversion ad phantasma'* of the Scholastic philosophers plays a dirty trick against them. It demonstrates not only that it is false that we should cast our eyes to fantasy in order to understand: it also demonstrates that we have to do without fantasy if we want to understand. For this distinction to be real, they *imagine* that the human spirit must be outside from God, and that God must be outside from the human spirit. This is obviously something absurd, because the spirit is nothing spatial from which we could predicate an 'inside' or an 'outside'. To say that something is outside from God is simply ludicrous.

Another possibility: They could be imagining that *besides* consciousness and intersubjectivity, spirits have a substratum, and that these substrata really distinguish themselves from each other despite that intersubjectivity builds some sort of drawbridge between them. We showed (III 4) that there is no *concept* of a substance distinct from its acts of self-consciousness and intersubjectivity. Therefore, what they have is an image. Every attempt of intellection fails if one does not start from this assumption: identity and distinction are intelligible, not imaginable. If they imagine the substrata, they become spatial beings, not spirits.

Or else: they place on the left side of the imaginary screen a 'time' in which God had not carried out the act of intersubjective appeal by which he made being the human spirit. We exposed extensively (III 6) that this time of physics is space, not time, and that these lapses in which there are no acts from the spirit do not exist.

The Hegelian commentator W. T. Stace incurred in the same mistake —only that this time, the mistake took place on the right side from the screen— when he affirmed that Hegel did not take literally the immortality of men but only as a symbol of "the absolute value of spiritual individuality. Immortality is a present quality of the spirit, not a future fact or event" (1955, 514). Stace pretends to correct both Hegel and the only adequate conception of time which is "something about movement" (Phys IV 219a 9), *in name of* an exterior absolute time that rules over God and the universe only because of the fact that we imagine time as a spatial line that goes from right to left, which exists independently from whatever may happen or not. As we have said, this line is space but no time; therefore, its right segment is no future, nor is the left segment the past in which the Scholastic placed God before he created the human spirit. Of course, the human spirit is not immortal if one understands by immortality that the spirit becomes coextensive with this line that goes to the right, for neither spirit nor true time have extension. Likewise, God is not coextensive with the left segment of this imaginary construction. Time is the '*abstractum* of transformation" (PR I 234), i.e. from the self-determination of spirit that transforms itself by giving to itself new and diverse determinations.

In general, when they say that God is or lies *beyond*, it becomes perfectly obvious that spatial imagination continues to play tricks with us. The true transcendence consists in the intersubjectivity we have discussed: it consists in that the self is always face to face with a thou which cannot be absorbed or reduced to the self. But this true distinction is at the same time identity. Only there the concept of distinct could have had its origin. If man were only surrounded by things and objects, he would never distinguish himself, just as his own body never distinguishes itself. We would have never come up with the concept of distinction. Consequently, he would not have the concept of a self either. Therefore (III, 7), the intersubjectivity which makes me a distinct being for the first time is the only possible way for a man to exist, that is to say, the only possible way to create man. But it is identity, for only in identity could there be distinction. We tie thereby a loose end that was (III, 8) missing.

Perhaps it will not be superfine to say that Thomas Aquinas also conceived creation in terms of identity. A posture to which a theologian of our century, Karl Rahner, adheres:

> Thomas does not regard the supreme mode of production of another, the creating activity of God, as an action that goes out from the agent towards and extrinsic patient, but rather as the immanent action of God, as God's free self fulfillment, which maintains himself completely within himself (De pot. q. 3 a, 15 corp). Thus, it becomes manifest that even the supreme causality towards the outside is more fully a *modus* of self-realization, and that transitory causality is a peculiar *modus* of formal causality (1964, 358).

They would not only have to charge Hegel with pantheism, but also Aquinas and the most serious part of the Catholic tradition.

We said (III, 8) that the only real causality we know and the only one of which we have a concept is the cause that determines itself. It follows from this that every attempt to understand creation that does not hold to the self-production of the spirit recourses inevitably to an imaginative figuration but not to a concept. Such attempts intend to play magic tricks by only pronouncing the word *cause*. Only intersubjectivity —as a simple and identical entity— has the capacity of producing something different without losing its identity. The Thomistic system as such is impotent to confront these things; matter and form are metaphors, evidently, but we are not dealing here with Literature.

And Hegel even makes a precision: "Creating is not a definite thought" (GP III 157). "The term 'creation' comes from religion; but it is a void word taken from imagination; in order to be thought and to have philosophical meaning, it must be more precisely defined" (GP III 240).

And let us see how Hegel himself defines it: "that we are not original and only in relation to God we come to being" (BS 239). Our third chapter exposed that it is the divine imperative which by means of intersubjectivity makes us come to be. "God is the manifestation of himself to the spirit, and this manifestation is to produce the spirit at the same time" (PR I 200). "This is the content of the Absolute: manifestation" (WL II 164).

The abstract intellect —which separates and stiffens everything— cannot understand that God is only conceivable in and by means of human intersubjectivity, whose essentially moral character we have highlighted. "To express it in more theological terms, the spirit of God is essentially in its collectivity; God is spirit in so far he is in his collectivity" (PR I 52). "God as spirit, if He remains on the other side, does not exist as vital spirit in his community, he remains unilaterally determined as mere object" (PR III 19).

8. Philosophy and faith

Since the theology of the abstract intellect does not realize that it does not have the concept of God and that it does not know what it is speaking about, since that theology does not realize that God is only conceivable in the moral imperative —whose obedience is a free act, despite the natural appearances— it blindfoldedly holds a difference between Theology and Philosophy, between faith and reason, that does not exist. When this theology speaks of "natural reason" in contrast with faith, it does not know what it says. Reason is not on the side of nature. Being rational is not natural to us. Although one demonstrates irrefutably with reasons —just as this book does— that the sensible data should not be trusted, man *can* anyway cling to the empirical data and distrust reasons; and that takes us to its naturality.

> Natural reason is a twisted expression. [...] Reason is rather not to be what one immediately is. Spirit consists precisely in raise above nature, to subtract oneself from the natural, not only being free from the natural, but to make in the natural that the natural obeys, adapts and surrenders itself to the spirit (PR II, I 10s).

In its most noble but infrequent sense, believing is to *know* something that is not in sensibility and that goes beyond what our senses witness. It is in this sense in which us the believers know —even by demonstration— that God exists (cfr PR I 85 n). The moral imperative, which is revealed by God, is knowledge of God with certainty. Evidently, in our context, the trivial meaning of the word 'believing' is irrelevant. "Believing is a word with many meanings" (PR III 199).

The noble sense alluded cannot be opposed to reason, because it coincides with the knowledge that is characteristic of reason, in contrast with the empirical knowledge of perception.

Perhaps the fundamental mistake of abstract-minded theology is to figure that one can know God doing without the moral imperative. Hegel is as intransigent as Kant in this point, and he is so with justified reason, since this is not only about the affective consequences: "the purpose and essence of every true religion, included ours, is the morality among men" (FS 105). It is the concept itself of God:

> True religion and religiosity stems only from ethicity, and it is ethicity that thinks about itself, that means to say, the ethicity that gains awareness of the free universality of its concrete essence. Only in the light of it the idea of God as a free spirit can be known; therefore, it is futile to look outside the ethical spirit for true religion and religiosity (EPW 552 A).

It is about the content itself of the idea of God. And all depends on which elements one takes as a point of departure to raise the mind up to God. "Kant fixed correctly the principal point of this elevation when he considered that the faith in God has its source in *practical reason*, for the point of departure implicitly encloses the content or material which constitutes the concept of the content of God" (ibid).

In the Kantian vocabulary, the practical reason is the one that perceives and obeys the moral imperative.

Kant pointed out with good enough reason that cosmological and teleological arguments demonstrated the existence of many things but not the existence of God. If one does not previously have the concept of God, the demonstration does not fulfill its purpose; and if one does have the concept, it is because we know God in the imperative without which we would not even be thinking, for the imperative is the one that makes us have a self and be humans. For that reason, Hegel told us: "It is very true that men have to be educated towards religion and not towards something that is not even there yet" (VG 128).

Our present chapter —which has dealt with the concept of infinite, and in which it was absolutely necessary to speak about our identity and distinction with regard to God— could not dismiss the droning thesis of the abstract intellect about the alleged difference between Philosophy and Theology, between faith and reason, whose fundamental mistake has just been pointed out by us. The moral imperative is not an empirical data; to know it is to have certainty of something that is not given by sensibility, and that is the noble sense of the word believing, but this knowledge is the essence itself of reason, the one that makes the man human, in contrast with the animals. Hegel holds exactly the contrary position than that expressed in the above mentioned thesis: "If theology is not philosophy, it does not know what it wants" (WG 838). "The difference between believing and knowing has become a common and vulgar contrast [...]. But in fact, in its essential content, this distinction is something inexistent" (VG 47).

Precisely the Trinity —which according to theologians is the most supernatural content there can be— is the key concept of identity and distinction, which is at the bottom of all *philosophical* problems, despite that superficial philosophy hast not yet realized this and cannot solve such problems.

> In fact, philosophy does no other thing than to understand this idea of Christianity (GP II 409). This content is present, as a teaching of the Christian church, in the Trinity. God is known as spirit only if one knows him

> as Trinitarian. This new principle is the hinge around which the universal history revolves. The history counts until and from there (WG 722).

The greatest vulgar and rationalist objection against the Trinity consists in denouncing it as a contradictory dogma since it affirms identity and distinction at the same time. Distinction in their persons —because there is nothing more distinct than a person—, and identity because the dogma says they are the only God, numerically one, so that these persons do not belong to a divine species. Evidently, the objectors think that they can give meaning to the words distinction and identity; but the huge surprise is that the only possible meaning of these terms is the one they have when their realization is the spirit, but the *true spirit is Trinitarian*. The first thing that the objectors missed is that the spirit is essentially intersubjectivity and distinction between persons.

> In regard of personhood, what is characteristic of the person, of the subject, is to depose its isolation. Precisely ethicity, love, and friendship consist in resigning the own peculiarity, the peculiar personality, and widening it until one transforms it into universality. [...] The trueness of personhood is acquiring it by submerging oneself in the other, by having submerged in the other (PR III 81).

In order to look with ontological and 'realist' disdain this identification with-the-other as if it were 'merely psychological', as if it were an emotional subjectivism that does not modify the real state of things, the skeptical would need to suppose that there is a *substratum* in each person that precisely underlies the acts of consciousness and remains unaffected by them. But we have showed (III 4) that such a substratum would be matter and no one would know what it consists in. If it is spirit, "the spirit is only real in regard to what he knows he is" (Rph 274), for "the spirit itself is only this perceiving itself" (GP I 93), "the spirit is not natural, it is only what he makes himself be" (GP II 494), "we know that there is only in spirit what he produces by his activity" (Ästh 166).

Personality does not consist in being eccentric. On the contrary, one is never so truly himself as when overcomes all peculiarities and individualisms and chooses for the good of everybody and is identified with then. Now, let us remember that the only possible meaning of the word *substance* is self-determination of the spirit, that is to say, the fact that the spirit makes itself what it is.

We said that the true spirit is Trinitarian. Self-consciousness consists precisely in that the spirit produces before itself *another* that *is itself* "because he has no figure of being another" (PR III 70 n.), and then the spirit suppresses it as another, for the recognition in it makes the other stop being another. The reincoroporation and recognition is the third element, but only in that moment the spirit is spirit, for there is no self-consciousness if there is no spirit. The spirit is that relation with oneself as another, regaining the other in the awareness of its identity with it. *Life* consists in this, taking this word in its most original sense, which is in fact the only possible sense as we will later see (V. 6). The abstract intellect kills all concepts; it makes them inert by separating and fixating them. It is obvious that what we have said in that paragraph is ideal and mental, but whoever opposes it to 'reality' still believes naively that he can define reality doing without the meaning this word has in reference to the spirit.

The spirit is originally Trinitarian: for that reason Hegel affirms that "God is the essence of man" (WG 575). This is what the Bible teaches us when it says that man was made in the image of God.

As Hegel says, the fact that philosophy does not anything more than understanding the Trinity is something that is not demonstrated because the concept of identity and distinction underlies all philosophical problems, but because the task of philosophy has been and still is to disentangle the true meaning of all concepts; however, all concepts find their original meaning in the spirit, and it happens to be that the spirit is Trinitarian. As an example of this, one should see the critique of Hegel against the system of Spinoza, a system which is the big synthesis of all the preexisting philosophies.

> The absolute substance is the true, but it is not the total true, the substance must be thought of as active, as living, and by those means to determine itself as spirit [...] The philosophy of Spinoza is only a stiff substance, it is not yet spirit, one is not with oneself. In that philosophy God is not spirit, because it is not Trinitarian (GP III 166).

It follows from this that Spinoza does not hit the mark with the meaning of concepts; his categories are simply "collected, without any further ado, he supposes him; no one knows how he arrived to them" (*ibid.* 167). Spinoza's demonstrations might be as geometrical as he wants them to be, but they are perfectly useless because neither he nor anyone knows the meaning of the propositions that are demonstrated like that.

Let us get back on track. If the Trinity is a philosophical matter, everything would indicate that only out of incompetence in their trade theologians could say that theology is not philosophy. It happens to them exactly the same thing that happened to Newton: they have not realized that they use concepts. Now, the task of philosophy is to find out the meaning of concepts, and we have seen (II, 5) how mistaken is the attempt of giving them meaning by decretory means. Even if theology would try to do without the 'spontaneous' philosophical sophisms —which are the sophisms of the epoch and of the layman—, even if it would want to reduce its work to the recital of Biblical passages, it is an undeniable fact that the Bible uses concepts, and hence it is impossible to understand its message without finding out the meaning of them.

"Theologians [...] ignore philosophy, but only in order that nobody gets in the way of their capricious reasoning" (EGP 156).

It is useless that they demonstrate us a thesis by saying that God revealed or leaning on the authority of the Church, if we do not understand what the thesis mean. Now, if we understand it, the result is that we do not only believe it but we *know* it.

One of the most impossible and contradictory acts is to say: "I believe in this although I do not understand it". To accept a thesis is not a physical act carried out by the hand; it is a mental act, and it consists in understanding the meaning of the thesis. It is impossible to hold a thesis as true if we do not understand it. We would not know what we take as true if we did not understand it. "To know is to have something as object before consciousness and be certain about it; and believing is exactly the same" (VG 47). The physical act of reciting with the mouth a thesis and saying that one believes in something is not *to believe* in the true of that statement; it is sheer exteriority; the fact that the performer does not have the intention of pretending does not transform that exteriority into true belief, for believing is an intellectual act, not a physical one. The content needs to be understood, for its *existing depends only in being understood*.

What meaning would it have to believe in anything?

Of course, the Trinity is a mystery, but "it is a mystery to the senses" (PR III 70). For the senses, one thing is *here* and the other is *there*, and they cannot be identified. However, for reason, for the spirit, the most understandable thing is its own reality, which is originally Trinitarian, as we have seen.

As we have said, the only true problem is to 'subtract oneself from the natural' and the sensible and become rational, i.e. spirit. "Without regeneration no one escapes from the sphere of the natural intellect to the speculative heights of the living concept" (BS 387). We said that it does not exist that what theologians called "natural reason", if we understand by natural that which can exist only by the forces of nature. *Man needs God to be man.*

Everything that God reveals to man has to be understood, this proposition is tautological.

The content of the revelation is not something physical, but an entity whose reality must be understood.

Now, a content to be intelligible needs concepts; and all concepts acquire their meaning in the realization of them called spirit. Only God can reveal us what spirit is.

Theologians would want certain knowledge to become contingent. But if this word has meaning, all knowledge is contingent, starting with the knowledge that we are spirit; to such a degree this is true that even nowadays theologians do not fully understand what spirit is (III, 1). The very same existence of the knowing man is contingent, if contingent has any meaning at all; therefore, all their knowledge is contingent.

They can define *contingent* as what is non-necessary, and as we have seen, they believe that they can define *necessary* independently from freedom and contingence (Cfr. III, 10) They are creating thus a pseudo-problem, as we will later see in detail (VI, 2 and VI 6).

"In absolute terms, theology is only what philosophy is" (GP III 64).

"God is the one and only object of philosophy" (PR I 30).

"Therefore, philosophy is theology, and dealing with it, or rather, in it is to worship God" (PR I 30).

"The spirit of truth is no other thing than the religious spirit" (WG 910s).

"Therefore, philosophy is not opposed to religion; what the former does is to understand the latter" (EGP 192).

"One has reproached philosophy that it wants to place itself above religion; but these is indeed false, since philosophy only has the same content and no other; however, it gives this content the form of thought; therefore, it places itself above the form of belief, but the content is exactly the same" (PR III 228s 2).

"Philosophy is not the only discipline that is essentially orthodox, but it is the most fundamental one; the statements have always been truth, the fundamental and essential truths of Christianity, are conserved and preserved by it [Philosophy]" (PR III 26s).

CHAPTER V

Logic and Natural Sciences

In the third and fourth chapters we exposed the core of Hegelian thought. Now, what has prevented people in our century from appreciating the Hegelian logic and the Hegelian critic of natural sciences has been the incomprehension of that core. That is the reason why I think we are ready now to understand the real range of that logic, which is the only logic possible and whose importance for human history increases every day.

1. Dialectics

The key of dialects is this: We know that the contradiction or antithesis between two terms *is solved,* and we know that because the meaning of each of them is the spirit, and the spirit is not contradictory.

The opposition or contradiction is due to the fixational intellect. Moving away from the concrete, this intellect converts the terms into many abstractions that are unintelligible because no one can give them meaning if they remain away from the tangible.

The engine of the dialectics is the exigency of understanding the terms. Once we reach the realization of these terms —which is no other than the spirit— they will become understood; and only then the oppositions and contradictions will be over. The pseudodialectics which posits that every synthesis becomes a thesis with a new antithesis and so on indefinitely, as we have seen (IV, I) does not share anything in common with Hegel and the dialectic; on the contrary, that is the indefinite progress that Hegel despises because it never achieves understanding. That dialectic cartoon lacks an engine, because it does not try to understand: it seeks *not* to understand. Therefore, every launch of a new thesis constitutes a display of arbitrariness.

> The abstract is false [...] The intellect resists the concrete; it insists on flattening it. By its reflections this intellect produces for the first time the abstract, the void, and clings to it in opposition to the truth [...] Philosophy is diametrically opposed to the abstract; Philosophy is precisely the struggle against the abstract, the permanent war against the reflection of the intellect (EGP 113).

"The abstractions correspond to the reflection of the intellect, not to Philosophy" (EGP 97).

"The treatment of such intellect consists in maintaining each determination or content of thought still" (GP III 262).

The pseudodialectic that tries to dissolve any particular notion and place it under skepticism is a cheap sophistic recourse, and this dialectic always stands in the middle of the road, since the end of the road is to understand.

"The dialectic that intends to dissolve the particular and to produce thereby the general, is not yet a true dialectic, it does not go yet in the true direction; it is a dialectic that is common to both Plato and the sophists, who were experts in dissolving the particular".

> The destiny of the supreme dialectics is to determine in itself the general, which has been the result of the mess of the particular, and in that universal to settle the contrasts, so that the dissolution of the contrast is the affirmative. Thus the universal is determined as that which dissolves the contradictions and contrasts, and all of this has made it in itself; therefore, it is determined as the concrete, as that which is concrete in itself. Therefore, it becomes determined as the concrete, as that which is concrete in itself. From that superior point of view, this dialectics is the properly the Platonic one. This is the speculative dialectic, that which does not end in a negative result (GP II 65).

We have already seen (IV, 3) that the true universal is the spirit itself; it is "the universal, but that which determines itself, that is concrete in itself" (GP II 68).

> The active universal, living, concrete, is which distinguishes itself in itself and remains in that free. This content consists in the identification of the one with itself in the other, in the many, in the distinct. Of all what is called Platonic philosophy this constitutes the true, the only thing true, the only interesting thing for knowledge; if one does not understand this, one does not understand what is fundamental (GP II 76).

The example of the real dialectic referred here by Hegel is the one we studied before (IV, 7): the abstract intellect separates identity and distinction and presents itself as incapable of defining and understanding them. On the other hand, when reason comprehends that the very spirit —which is essentially intersubjetive— is the only possible realization of both the concepts of identity and distinction, it realizes not only that they do not contradict each other, but that they have to be identical in order to exist and have meaning. Hence the opposition between these two concepts was due to the intellect kept them apart, and converted them thereby into abstractions, since they are not separated in the concrete; they are identical and only in this way can they be understood. Evidently, for the spirit, there cannot be something as concrete as itself.

"If the beginning was the universal, the result is the individual concrete, *subject*" (WL II 499).

"Sense and meaning are something concrete" (GP II 591).

"This second negative thing to which we have arrived, the negative of the negative, is the suppression of contradiction; however, [...] it is not the product of extrinsic reflection, but rather the most interior and objective element of life and spirit, that by means of which a subject is a person and is free" (WL II 496s), "the pure personality, the most subjective thing there is" (WL II 502).

One cannot express with greater explicitness that the real dialectic is an exclusive patrimony of the spirit: the only solution of antinomies is the most intimate element of the life of the spirit. The pseudodialectic of materialism is an anti-scientific whim, because, being the realization called spirit the only one that makes that the concepts have meaning and become, for the first time understandable, such pseudodialectic is not directed to *understand* and hence it has no justification in its transition from one concept to other.

"But as absolute negativity, the negative element of the absolute mediation is the unity which consists in subjectivity and soul" (WL II 497).

The commentators —specially the Marxist ones— considered that dialectic was important by itself, but the really important thing for Hegel is to show that all the concepts mean spirit. Actually Hegel laughs at the "dull set square of triplicty" by which Kant "inserted the thesis, the antithesis, and the synthesis carelessly" (GP III 385).

It is crucial to distinguish clearly between the Kantian and the Hegelian solution of the antinomies. Kant claimed that human mind is incapable of knowing the reality; hence, the mental categories do not say us anything about the real. It is then irrelevant that they contradict each other, e.g. that free will contradicts the necessary, finite contradicts the infinite, simple contradicts the compound, etcetera. It would be bad, Kant says, that the reality itself was contradictory; but it is not, because we do not know anything about reality. Our concepts don't reach it. It is the *incognitum* x. Our concepts contradict each other, but reality remains immune from all of this.

The big question that Hegel raises up against such a solution is: where do these concepts come from? How can one explain their existence? And the great objection: the concepts are still as contradictory as before, the antinomy has not been solved; it has just been reinserted into the subject. Kant is very tender with the reality: he does not want it to be contradictory, so instead he makes the subject contradictory.

Hegel, however, confronts the problem directly: *the contradiction has to be solved*. That it is really solved is something that our above mentioned example of identity and distinction proves.

> Since each of the two opposed sides contains its other within itself and neither can be thought without the other, it follows that neither of these determinations, taken alone, has truth; this belongs only to their unity. This is the true dialectical consideration of them and also the true result. (WL I 191)

"The third is what is properly speculative, that means, to know the opposites in their truth are one." (NH 415).

Now, the dialectic of identity and distinction underlies the dialectic of the simple and the compound, of the continuous and the discrete, the qualitative and the quantitative, and the extensive and intensive.

For the simple and the compound we need to examine a larger quote:

> The Eleats said that God was the one, which means to say, the simple. Now, the one, and the simple as such is completely void. What follows is that the simple can also be concrete and at the same time multiple in itself, so that that despite the distinction the unity remains. There is an example of this in our spirit. When one says, I am one, I am simple, I am a point, or I am one with myself, one has a representation of the self as something perfectly simple; there is nothing as simple as that self. However, we know that that self is a world in itself. Each man is in itself a world (the entire world); in this simplicity each man is an abyss that encloses an infinite content.
>
> By thinking of oneself or remembering things, one brings into light the richness that one possesses as such. The self is, therefore, entirely simply, and at the same time it possesses a whole richness in itself. And when we say spirit instead of self, here we have from the very beginning not the representation of an abstract, but rather of a living organism. The spirit must be a totality, and yet this totality must be as simple as the self (EGP 277).

Let us comment the above quoted text. The concept of simplicity, evidently, does not have an empirical origin, because everything we can point with our finger, is always compound, in the sense that if we wish, we could divide it, at least imaginarily "one only points out matter which is compound" (GP I 359). Hence, only self-consciousness could make the idea that something simple exists. And actually, as we said before, the self is the simplest thing there can be; the very identity of the self depends on that. It is very important to realize that the idea of simplicity is something positive; it is an idea which has its proper content; it is impossible to define simplicity by denial of the compound, because then we would have to define the compound as that which is not simple, and by this circularity we would not obtain any content at all; we would not understand anything when we refer to the simple and the compound, but that is something that contradicts everyday experience.

We would have never called anything compound if it was not in comparison with something that is directly simple. In fact, the constitutive simplicity of what we call 'self' is so essential and prominent that in comparison to it any other thing hardly deserves to be called simple. Obviously the subtle thought that nothingness is simple could not be in the concept we are dealing with; it is something derived, because we do not understand nothingness by empiricity or by reflection; nothingness is a refined abstraction which is constructed through many negations of the positive; it cannot be the origin of any concept at all. Nothingness is not simple or compound: it simply does not exist.

Despite that the self is the simplicity par excellence, it is evident that the same self, which consists in self-consciousness, is constituted by a multitude of intellections, volitions, wishes, memories, decisions, etcetera. Without that multiplicity we would not have the idea of calling the self simple, of saying that it is identical to itself and that is why, it is simple; the very idea that it identifies with itself implies some multiplicity of one and the same self. In order to be simple it needs to be compound. And vice versa too: those multiple life experiences, if they did not constitute the one and only identical self, they could not be called elements of a compound, but everyone would exist by itself and with no relation to the rest. The abstract intellect does not understand either simplicity or composition, because it incurs into the absurd "commits the same absurdity of making that which is pure relation into something devoid of all relation (WL I 211)".

The contradiction or antinomy is solved when we prove that *the concrete* to which the concept of simplicity refers is the same reality to which the concept of composition refers, and hence, the existence of antinomy has its origin in the stubbornness of the abstractions of the intellect and they want to remain separated and to still be abstractions. This situation makes it impossible to give them an intelligible meaning.

Now, the antinomy of the continuous and the discreet is hereby resolved. The continuity of the self that we called simplicity does not only exclude the complicity of the diverse life experiences, but identifies with it in such level that the former cannot be defined without the latter. And *vice versa* too: no reality cannot be called discontinuous if its elements existed each by itself and without continuity between them; they would be separated entities with no relation whatsoever between them. They would not constitute the discontinuity of one only and the same entity.

Continuity is defined thus: one that simultaneously is many. The definition of discontinuity or discretion is: many that simultaneously are one. Hence, they are the same. Scientists only discover a truism when they affirm that continuity and discretion are complementary concepts (as we will later see, this is what Bohr's complementarity principle says). Hegel said it one century before: "none of these determinations, taken separately has truth, but only the union of both" (WL I 191). The physics would have saved one century and many headaches if, as the real science requires, they had demanded rigorous definitions of every term they use.

"The continuity is this content of the *identity-with itself* of what is discrete; the follow up of the distinct ones in those that are distinct to them. Therefore, magnitude has in continuity the element of *discretion* immediately" (WL I 180).

The pseudodiscreet of the imagination consists in many which are not one: each one is entirely on their own. The pseudocontinuous of the imagination e.g. a line, the stretches are contiguous, but they lack continuity, because each of them can and exists on its own.

As we have seen, the reason why the physic and the natural scientist did not demand the definition of their terms is because they thought that meaning was some empirical data. We have proved that it is an illusion, and even in order to know if the empiric data corresponds to the belonging word it is still necessary to define it before, since the possession of any empiric data does not exempt them from the obligation of defining. In this particular case, a little bit of reflection would have been enough to understand that sensation could only justify contiguity by itself, and besides, this is not the same as continuity. The continuity is something very metaphysical; there is nothing similar in the physical. In the best case scenario, what quantum physics was looking for was the juxtaposition of points, but juxtaposition is not continuity.

The sensibility ignores if the many elements that it thinks it verifies are one or many, i.e. does not know if they constitute or not only one individual, because, as we saw (IV, 6), individuality and sameness are not an empirical data. Hence sensibility cannot witness continuity. Further, its testimony of contiguity is completely unreliable for science, because it constitutes one single macroscopic impression due to the peculiarity of our sensorial organs. The atomic and corpuscular theory repudiated that testimony from the very beginning.

We said that Bohr's principle affirms complementarity between continuity and discretion. We need to support our affirmation because one commonly thinks it is a complementarity between wave and particle. Margenau saw that this common interpretation is a mistake. "The point is that we have a reasonably satisfactory theory of the electron which assigns to it neither the carácter of a particle nor that of a wave." (1978, 89 n.13) D'Abro has understood this with all clarity too:

> According to our earlier views, the waves, being mere probability waves, must be regarded as symbolic, whereas the particles represent physical reality. But we must now suppose that particles also should be regarded as symbolic, for they come into existence only when a position observation is performed. (1952, 652).

Schrödinger himself, whose equation is the operation center of quantum theory, emphatically told us that it was a mistake to conserve the concepts of classical physics (Maxwell included) and not just denied them accurate definability. This is literally the way he puts it: "Concepts themselves must be abandoned, not only their clear definability" (1934, 519). That one still keeps talking about waves and particles shows that neither the public opinion nor physics themselves dare to accept the *facts* that quantum experimentation has discovered. The complementarity is not between wave and particle.

But if we take it part by part, the issue turns itself pretty obvious. One cannot talk about particle or corpuscle if the precise localization, the precise trajectory and the precise mass are denied. On the other hand, if Physics talks about the de Broglie's wave in association with the electron, it is evident that it does not consider the electron as a wave. In fact, the vector describes a global property of a physical system; it does not describe in fact a wave or a particle. Although it is common to call Schrödinger's *psi* ondulatory, we are specifically warned that such a wave is a complex function and gives us plausible information about the whereabouts of the electron, but according to what we have just said, the electron is neither really a particle nor a corpuscle. Bohr himself emphasizes this when he speaks of "the renunciation of the absolute significance of conventional attributes of objects" (1958, 64).

The complementarity between continuity and discretion —affirmed by quantum physics— is an overdue and not very lucid acknowledgement of the fact that none of these two concepts is true taken alone, but only the union of both. That was the Hegelian solution of the respective antinomy, but in order to provide that solution the only thing that Hegel needed was the meaning of the concepts, or in other words, the fact that they can only have a meaning if they are united, and the realization of both can only happen in the spirit.

"We call dialectics, the supreme movement of reason in which apparently and absolutely separated things go from one another by themselves, by what they themselves are, and the supposition that they are separated is thereby suppressed." (WL I 92).

"The watch over the categories employed by the intellect that insults Philosophy is what the latter needs" (BS 370s).

In the case of all the examples we have just said, it would be redundant to say that the contradiction is effectively solved and does not remain as contradictory as before (this was the case in Kant). Likewise, it would also be redundant to repeat in the light of each example that the synthesis and the solution are not the beginning of a new antinomy, but rather their final solution, since one can only *understand* in it the meaning of the concepts involved.

The last quotation says very well that we have not put enough attention to the categories. If physicists and philosophers would have understood that continuity is one that is many, and that discretion is many that are one, they would have realized that they are the same, and that the fake security of the man who separates them and thinks he understands them is only imagination. The common sense has a lot of abstract intellect, which —as we have seen— thinks like the imagination. Hegel says: "common sense, just as the unilateral abstraction, tends to invoke itself" (NH 166).

The same happens with the contrast between the qualitative and the quantitative. Shamelessly, one supposes that they are diametrically opposed and that everybody knows what they mean but when the positivist logicians try to clarify what a 'qualitative predicate' is, they sink into failure. They say that a predicate is qualitative if it is not applied to a finite number of objects. By excluding the finite number they think they are excluding the quantitative. But undoubtedly the number of green objects in the world is finite, and yet 'green' is one of the most frequent examples of qualitative predicated. In addition, finite and infinite do not have any empirical meaning (cfr. IV, 1).

As for the scientists, it is obvious that the negligence with definitions can have its origin in the belief according to which it is enough to point with the finger at some empiric data to give meaning to the words qualitative and quantitative. But it is still more obvious that if we do not know beforehand what 'qualitative' means, we will never know on what aspect of the empirical fact we have to focus our attention. The color green gets in by the senses, but that the color green is a quality does not get in by the senses. That the green stops being green when it loses its essential property, is an entire metaphysical detour that the *intellect* promotes without the slightest intervention of the empirical data.

We have said (II, 6): even if the object is really contemporaneous, that is not enough for the contemporaneousness to be an empirical data, not even for seeing the object the idea of contemporaneous will take form in the mind, in the same way, even if the object or fact would be qualitative in reality, you can not conclude his qualitative character as an empirical data.

That the quantitative character is not an empiric data was demonstrated before (IV, 1 *fine*): the sensibility does not perceive the number. We beg you to confer.

This conclusion is imposed on us: these two concepts were originated not in the sensation but in the reflection of the subject on itself, and there cannot be contradiction because the meaning of both is the spirit; it happens that the abstract intellect separates both notions in the imagination and stiffens them, making them unintelligible. Quality is a modified characteristic; when it is modified, the entity is not longer what it was, i.e. it loses its identity. Quantity is a modified characteristic in which the entity keeps being what it was, i.e. it conserves its identity. These couple of definitions —which are, by the way, the only ones available— show that they are concepts which are openly anti-empirical, because the concept of identity is present in both (cfr. IV, 6).

From that definition, we conclude that the quality is a limit and implies denial; beyond that characteristic the entity is not anymore, quality is the "negative itself" (WL II 109).

On the contrary, the quantity is "the limit which is equally no limit" (WL I 332), "the *indifferent* determinateness, i.e., *posited* as *sublated*" (WL I 331s), because the entity affected by a quantity is still what it was if this is modified; the entity is beyond this characteristic.

Now, for the realization of that notion of quality, one needs a substance without substratum. It has to be something that, regardless of the qualitative affection, remains the same. In other words, the quality has to reside in a substance whose accidents are not accidental to it. The third chapter (III, 4) showed that this substance is the spirit. The determinations that the spirit acquires in its lifetime affect it completely, it identifies with each of them; there is no substratum that remains unaltered; the spirit is its own acts.

At the same time the fourth chapter (cfr. IV, 3), showed that the determinations that the spirit gives to itself are indifferent; those are limits that are there to be suppressed. With Hegel we said: "Infinitude [...] is to pose an immanent and proper determination, which is something indifferent to the one that poses it (NH 87s)". That is why we can say:

"The determination of pure quantity belongs also to the ego" for the self is "the continuity of universality or being-with-self un-interrupted by the infinitely manifold limits, by the content of sensations, intuitions, and so forth." (WL.I 182).

This is why Hegel can conclude: "quantity is the truth of quality itself" (WL I 333). What seemed to be limit turns out to be null within the spirit, which is the only realization that goes deep down into the bottom of the entity.

Looking at things more closely, we see that the quality is the discontinuity, while the quantity is the continuity, and we saw that continuity and discontinuity are identical. In the same way, its multiple qualitative characteristics make that spirit a compound entity; and the nullity or indifferent character of such limits is what makes the spirit simple. We are always in the same dialectic of identity and distinction.

All this could only be denied by someone capable of assigning to the concept of qualitative an empirical meaning, a meaning different than the one contained in the alluded definition; but nobody is capable of doing that. It should be evident that there is nothing as qualitative as the spirit's characteristics, and that the origin of the concept of quality is self-awareness. The idea as whether something can stop being what it is could have only been formulated around the unmistakable identity of the spirit. There is content: the individuality, which has in the spirit its true meaning. On the contrary, to call qualitative something material is always a tautology that says nothing. Let us think in this hollow inanity: the green, if it is not longer green, is no longer what it was. Evidently, this is an extrapolation: we are projecting towards the outside of the spirit's unmistakable identity, as if this could have in the material some meaning.

It should be evident that there is nothing as qualitative as the spirit's characteristics. And yet, the spirit dominates the limitations (every determination is limitation) and is still the same not just in spite of them, but in and because of them, because the identity only has sense between one 'same' in certain character and that 'same' in other character. Hence, they are quantitative for if they change it keeps being what it is. And the truth of quality is quantity.

Let us move to another example of antinomy that is shorter but no less interesting: the intensive and the extensive. This contrast is not the same between continuous and discrete. The continuous or discrete resides in the very quantity and lack of limit. The extensive or intensive drives our attention to the limit, to the determination of the stub or extremity, which allows us to speak about the quantity (amount) that is not important for the continuity or discretion because a magnitude is continuous or discrete before reaching the limit although this may never be reached.

The reader will agree with us in that common people talk about the extensive in contrast to the intensive with a sensational ease, as if the definition was not necessary because 'everybody' knows what these concepts mean everybody knows what are opposed to.

Supposedly, the elements that constitute the extensive magnitude are exterior to each other: if we say one hundred meters, every meter stands apart from the other ninety-nine meters. In opposition, the intensive magnitude pretends that the last element, the stub or extremity —ordinarily called degree— is the only one that exists, and the others are absorbed by it. If we say 'forty degrees', one supposes that the twelve and the thirty-five degrees do not exist, or that they have been absorbed by the forty degrees. The extensive magnitude is multiplicity, while the intensive magnitude, apparently, does not have multiplicity in itself but is rather a simple determination. It is easy to see that we are in the same dialectic of the one and the multiple, the simple and the compound. The origin has to be the subject's reflection over himself, but this time we will not insist on it; it is enough to show that the antinomy or contraposition is more imaginary than real.

The mere fact that the sound of a high tone, which is intensive, consists in a bigger number of vibrations per second, which is extensive, raises enough suspicion. In the same manner, when the physic talks about a higher degree of temperature, he actually refers to the height of the mercury's column, which is extensive. The suspicions are confirmed if we analyze the concepts.

Even if we talk about an extensive magnitude, each number we pronounce absorbs the numbers smaller in the same sense in which the degree supposedly absorbs the smaller degrees. By mentioning a small extensive magnitude —five meters, for instance—, it is possible that the *imagination* will figure that it is seeing the four extended meters just besides the fifth one, but if I say 2,728,585 meters, it is false that the fantasy imagines all those meters extended out along each other; they have been absorbed by the biggest, even though we are dealing with an extensive magnitude. The decisive factor should be the concept and not an imaginary figuration that can be present or not.

On the other hand, the absorption of the intensive is not a real absorption. When I say forty degrees, the word forty supposes that thirty-nine is distinct from forty, and that thirty-eight is distinct from thirty-nine, and the same until we reach the unit, but it means that they are outside from each other, that they are external to each other; without that implicit journey, the forty could not represent an amount or quantity, but only an element that exists by its own. In order to signify a determined quantity, it is necessary that the thirty-nine is something distinct, and that would apply to the thirty-eight and so on indefinitely. We cannot say that they are exterminated or absorbed; each one maintains its own different meaning and exteriority in relation to the others. What appeared to be simple is, in fact, not the case: "as indifferent to the differently determined intensities it has within itself the externality of the amount; and so intensive magnitude is equally essentially an extensive magnitude." (WL. I 217)

After Hegel's death, Physics has made a big fuzz about the distinction between heat and temperature. But even assuming that this distinction is real, what is truly necessary is to re-think if that coincides with the distinction between extensive and intensive. This is a typical case in which the scientists, without examining the meanings, use the first couple of existing words for two entities that seem distinct to them. Furthermore, it is possible that we are not dealing in this case with realities which are empirically verifiable, but with one reality that has been postulated as subjacent in order to *explain* the existence of the other; and it is possible that both have been postulated in order to explain the empirical phenomenon of the varying height of the mercury column. Soon we will see the Hegelian critique against those supposedly explanatory entities.

As a last example of true working dialectic let us mention the *effective* solution to one of the most seemingly irreducible and unsolvable antinomies, namely, the antinomy between necessity and freedom, with which even Kant did not know what to do. About it Hegel says:

> Or take an example from a higher sphere: We say that man has *freedom*; the opposite content is *necessity*. 'If the spirit is free, then it is not subjected to necessity': and then the *oppositum*: 'His will, thought etcetera is determined by necessity, hence he is not free'; and they say: 'one excludes the other'. In such opinions the contents are taken as mutually excluding, as if they did not constitute something concrete. But what is true is the union of the two opposites; we have to say that the spirit is free in its necessity and only therein has freedom [...] What is easier than to fix the concrete is always to say that necessity excludes liberty and vice versa (EGP 116).

We indicated (III, 10) that nobody can give an empirical meaning to the word necessity. Hence, the origin of this concept can only be the reflection of the subject on itself. But the only necessity that we know by reflection and self-awareness is the imperative imposed to us whether we like it or not. That moral imperative, however, is the one that makes us free for the first time by providing us a responsibility that generates self-awareness; only in the moment of being addressed by obligation and in the response to that obligation man becomes truly free. Before that, he was only an animal. "In duty the individual frees itself for substantial freedom" (Rph 149). "Like ethicity, *true* freedom consists in that the will does not have subjective and egoistic ends but only those of universal content" (EPW 469 A).

Not only due to its terrifying practical consequences, but also from the rigorously analytic and theoretical point of view, it has been a giant mistake to conceive liberty as a negative issue, as the lack of something. One deducts from this that man is free by nature, that all the primitive human groups are free since the beginnings, for if freedom does not consist in something positive that has to be acquired, man posses it by the mere fact of existing, without the need of conferring himself anything. One sees in that conclusion how big such a mistake is, for man evidently is not born free; during his first years he lacks freedom in the exact way little animals do. In regard to this point of unequalled importance we could not present a testimony more unexpected than that of Rousseau:

> What is meant by a virtuous man? He who can conquer his affections; for then he follows his reason, his conscience; he does his duty; he is his own master and nothing can turn him from the right way. *So far you have had only the semblance of liberty*, the precarious liberty of the slave who has not received his orders. *Now is the time for real freedom*; learn to be your own

master; control your heart, my Emile, and you will be virtuous (Emile, V: the italics are added).

That this is said by the greatest panegyrist of the natural man is something the reader should not disregard.

I repeat: if liberty consists in not having, if liberty is a form of nothing, anybody has it. It is easy for the natural man to come to the world equipped with something that consists of nothing. In fact, things are the other way around: "freedom has to be essentially affirmative" (WG 775).

To define liberty the way Hobbes does, as the absence of impediments, is totally circular and lacks content. Hobbes pretends to give the word liberty a physical and empirical sense, and evidently confuses 'being free' with 'being loose', like if the wild pig in the jungle was free just because he can go anywhere. But the word impediment does not have the empirical meaning that Hobbes would like it to have. If it is a *current* physical motion, it is indeed an empirical data the fact that a jail impedes it; but then one should affirm that I am free in so far I physically and effectively move myself (and nobody sets impediments on me). If I deliberatively decide to meditate at ease about my issues, Hobbes has to affirm that I am not free because empirically, there is no motion without obstacles . Surely what Hobbes wanted to say is that I am free if I *can* move myself without obstacles, even if I do not actually move myself. What he would say is that freedom consists in the capacity of being able to move oneself. But we have made evident (III, 9) that 'can' and 'being able' are not empirical data; the only way to give meaning to those words is by means of self-consciousness, and the meaning that we perceive there is not a mere possibility or an incompatibility of terms, but a positive capacity: the power of the spirit of giving itself different determinations. The only meaning possible for the word *can* is self-determination of the subject, the free will that Hobbes wanted to put sideways by means of his attempt of configuring a physical meaning.

In addition, it is obvious that even in political life the lack of physical impediments would be useless if the subjects were not psychologically free. Huxley and Orwell have showed this clearly. If the victim is drugged, the captors do not need chains or fences. And more efficient than drugs is the systematic disappearance of knowledge in education, culture and information, which means to say, the systematic disappearance of concepts. Man is free insofar he thinks, insofar he has distinct points of reference than those monoidetic stimuli that irremissibly motivate the animal. Only he who has *awareness* is free.

The absence of physical impediments only has sense in function of the psychical liberty and it depends on that, for only in function of the second is possible to give meaning to the first.

If freedom is not something negative, if it does not consist of a lack, Rousseau's quote acquires an extraordinary relevance: going against his dogmatic apriorism of a natural goodness —which made him famous—, Rousseau honestly recognizes that freedom consists in the genuinely moral act that overcomes the natural tendencies and acts even against them. The demagogy that makes fun of us in front of the great audience by ridiculing 'monastic' asceticism and praising a 'happiness' that as we will see (VI, 2), no one can define is out of the question. That sort of people can remain with their triumph. What we are dealing here with is truth. Kant demonstrated the trueness of Rousseau's intuition. And Hegel sums up concisely the Kantian analysis:

> Formerly, the so-called theory of happiness prevailed in the practical; morality was grounded in impulses; the concept of man and the way in which he had to realize his own concept was conceived as happiness, as satisfaction of his impulses. Kant demonstrated correctly that such thing is heteronomy, not autonomy of reason; being determined by nature is not freedom (GP III 334).

In the cases in which the course of action is decided not by the self but by an extern impulse one cannot speak of autonomy or freedom: the action is heteronomous.

"Man is not autonomous because movement begins in him, but rather because he has the power of stopping the movement and thereby to break his own immediacy and naturalness" (VG 57).

The spirit is what *he* makes of himself, not what impulses make him be. This is why we said (III, 2) that "liberty is the only true thing in the spirit" (VG 55), and that: "the only determination of the spirit in which all the other ones are contained is his freedom" (NH 58). In the first chapter we pointed out: "The natural is rather what the spirit must suppress" (GP II 107).

The necessity that is produced by the moral imperative identifies itself completely with freedom. And only in that moment is understood the meaning of necessity and liberty. The abstract intellect was the one that separated and isolated them, and in doing that it made them unintelligible; that abstraction was the cause of the antinomy; in concrete, that is to say, in the spirit, these two concepts do not contradict each other; on the contrary, they have the same meaning.

2. Logic

We had to multiply the examples of real dialectic in rigorous analysis of the terms, because both scientists and superficial commentators believe that in Hegel the transition from one concept to another is capricious and of a literary kind. For Hegel, on the contrary, "all logic consists in this" (WL II 495).

The logic consists in the 'it follows'. Everything else is accessory. Logic consists in this question: With what right is one concept *inferred* from another? With what right can one *deduce* a proposition in which the subject has a predicate another than the one he originally had before? How can one justify the transition from one predicate to another, which is evidently a transition from one concept to another?

In the light of the above, it is enormously disquieting that scientists and philosophers do not realize the total revolution that Hegel introduced in this science, upon which the scientificity of all sciences depend, in spite of the fact that he specifically warns that the Aristotelian logic is useless and that he, Hegel, is founding the real logic:

> The syllogism in the fashion of the abstract intellect —which is the one rendered by the traditional logic form— has the meaning of a content joined with a different content. In contrast, the syllogism of reason has the content that the subject, etcetera, is joined to himself; the syllogism of reason is that some content, God etcetera, is united to Himself by means of distinguishing itself. This identity constitutes the essential element of the speculative content, of nature's syllogism of reason. Aristotle is, therefore, the founder of traditional logic; his forms correspond only to the relation between finite to finite; truth cannot be understood under such forms (GP II 241).

The real logic is based on what this chapter has underlined: one concept 'follows' from another *because they are not understood separately.* The logical necessity is the exigency that a concept has meaning and is understood.

The implication between the concepts is what traditional logic as well as set logic ignore. But that implication is what allows a real inference of something new, of something that effectively increases our knowledge and is not tautologically reduced to what was said: what was said does not 'follow' because we had it already. For instance, the well-known syllogism: 'All men are mortal, Socrates is a man, therefore, Socrates is mortal', does not teach us anything in the conclusion, for what it said was already stated in the first premise. The first premise could have not been stated had we not known that the content of the conclusion was true. Such reasoning does not make us wiser.

In contrast, the logic of Hegel begins with the concept of being, and because of the exigency of providing this term with a meaning, it leads us (cfr III, 3) to a new content: the activity of giving determinations to oneself. Being and existing can only consist in that. That really does increase our knowledge. The inference can be denied only by he who is able to provide the word being with another meaning, but we saw already that such thing is impossible.

We saw (III, 4) what happens with the initial concept of substance: it is not understandable that something can exist in itself if it doesn't give to itself the determinations of its existence; and then pass from the substantial content to the subjectivity content.

Tautology is the way-out of the abstract intellect that does not understand. When it comes to *understand* the concepts, our knowledge actually increases.

The only path that fixational logics can go through in order to avoid becoming sheer tautologies is the hypothetical syllogism. I call here hypothetic not only the syllogism that has that explicit form, but also every syllogism in which one of the premises is *taken* as truth and is not demonstrated. This occurs extremely often, for the common logician thinks that, if every premise must be demonstrated, the syllogism contained in the demonstration would also need premises that need to be demonstrated, and so on indefinitely. When that happens the common logician takes a secure path and *assumes* some of the premises.

The point has a pivotal importance for the very concept of scientificity. Science is concrete and demonstrated knowledge; otherwise, one can 'choose' to accept it, just like the worldviews which are continuously advanced by pasquinade writers. Although the individual scientists is allowed to suppose some things because the life of a person has some limitations of time and energy, science as such cannot suppose anything;

its obligation is to demonstrate. But if common logic replies that the premises of its syllogisms should not be demonstrated, what they are in fact telling us is that science does not exist, or, if anything, a very funny and easy thing to do. One can embrace any thesis, as arbitrary as it may be, for one does not need to demonstrate the premises that ground such thesis.

The traditional syllogism is a deceit. It simulates to demonstrate the propositions, but in the end it leaves them as uncertain as before, because its truth depends on the truth of the premises, precisely that which is left aside.

One can make a big mess which is apparently true because it is expressed in conditional terms that are formally valid. But knowledge does not advance a single step by this procedure, since we do not know if the conditions are met or not. It would be a mental game with no relationship to reality and truth.

How was it possible that essayists and the users of traditional logic did not realize that, if premises are uncertain, concepts themselves are also so.

Indeed, one can build up a perfectly correct syllogism in which all the concepts are uncertain. In that case, we may ask: Where are they taken from? How can we know their meaning? And more importantly: How can we know that they correspond to reality?

The desire of obtaining some judgments by *empirical* experiences is, of course, an illusion. The judgment is never 'given'. The allegedly empirical judgment does not identify itself with the empirical data. Despite that they call it observational, the allegedly empirical presupposition consists in words, while seeing or touching does not consist in words but in shudders of the diverse parts of the nervous system. The basic and observational propositions is only an *interpretation* of sensible data, highly questionable because it is constituted by concepts, and no concept as such has empirical meaning, as we have repeatedly demonstrated. In order that logic *starts* working "what is perceived must have at least the meaning of a universal" (PG 185); and in fact "for us the object can be nothing else but our notions of it." (WL I 15); "the syllogisms with which one demonstrates are concepts formed by the subject" (GP III 280).

It is then when, with all its virulence, the problem of demonstrating premises indefinitely is posited, for it depends on the concepts. *Some* certainty has to be the condition of possibility of science; otherwise, one would have to define science as the 'uncertain', and there would be no distinction between it and ordinary knowledge, and hence science would not exist.

Let us repeat the question: how do we know if concepts capture reality or not? It is evident that the famous 'correspondence between mind and reality' has to consist in the 'identity between the mind and reality'. Hegel says: "Only in thought exists the true correspondence between the objective and the subjective: *that is myself*" (GP II 165).

There cannot be any separation between the spirit and the concept of spirit; there can be no distinction there, and in fact, there is not any. We saw before (III, 1) that the spirit *consists* in knowing itself. "The first thing one has to do when it comes to the concept is to quit believing that the concept is something that we have, something that we have in ourselves [...] What we call soul is the concept; the concept as such becomes existing and is the spirit, the self-consciousness" (PR I 220).

In *this is myself* there is certainty as whether the concept adapts to reality or not, as whether the concept captures reality or not. What we have expressed in our previous chapters demonstrates that all concepts stem from the concept of the *spirit*. As Hegel says, "the knower [...] has in its concept the whole essentiality of the objective world" (WL II 438). We have demonstrated that it is an illusion to define reality independently from the subject (II, 7), and that 'out of thought' is an expression lack of meaning.

The relations and interferences between concepts —in which logic consist— are only then possible because all of them point towards the same reality: the spirit. Besides, that is what judgments affirm: that one *is* the other. Without Hegel's contribution, logic today faces a dead end: it deals with tautologies or with unjustifiable synthetic judgments.

By way of appendix, one should notice that set theory is a desperate attempt of evading the concept, by making everything visual, spatial and imaginary. It figures that, if an object is a part of a certain set of objects, and that set is the part of a larger set, it is a logically justified step to affirm that the object in the first set is also a part of the second one. It figures that in this 'it follows' one does not employ any concept whatsoever. But it is primal to state that —despite the opposite figurations—, when the set logician says 'the set of metallic objects' he is neither pointing out with his fingers to all the metallic objects, nor is he imagining them, nor he manages that we imagine them. Consequently, there is no such set. If we understand him it is because we both *have the concept* of a metallic object. The extentionalist does not have before him the set of metallic objects in imagination, in empirical perception or in the mind: the only thing that he has before him are concepts. And to affirm that a certain material object forms part of a concept is preposterous.

How can one know whether x belongs to set A, when we do not understand the word by which one is trying to encompass set A? It is not true that someone can designate by means of this word certain set of objects, for we do not know to which object —among the many ones there are in the world— this word is referred. To designate is not a relation that exists between the word and the belonging objects without the intervention of the knowing subject.

Since the set logicist does not have before him either set A or B, if he is truly referring to them it is due to the fact that he has in mind the two belonging universal concepts, and the relationship between the both sets of which he speaks is a relation between concepts. Now, there can be no other thing between concepts as relations of content; any other relation between them would be a mere metaphor. But the relations of content are precisely those which are studied by the Hegelian logic.

3. Force

In the fourth part of this present chapter we will show the most pervasive critique that has ever been done to the natural sciences, which is contained in the work titled *The Science of Logic*. Natural science disciplines had not attended to such devastating critique, not even after Einstein and Quantum theory, although in some way, both Relativity and Quantum physics confirm what Hegel said. Nevertheless, it happens that the intellection of the above mentioned work depends on understanding the message from, what Hegel calls, the second book, the essence book, in which the concept of force is so paradigmatic that deserves to be examined by us in full detail. That second book is the one which contains the explicative concepts, in contrast to the merely descriptive ones; but among the explicative ones, force, according to Hegel, is 'the most notable one' (GP III 84), while Hegel warns that it is not actually a concept but a certain 'mental way', certain 'way of thinking'.

It is true that General Relativity with its realm wants to delete from the world of Physics the gravitational force which is the key to the great Newtonian construction. During the eighties many physicists, following Einstein's example, pretended to delete *every* force by spaces formed up to eleven dimensions. But, in the first place, we will soon see that the Hegelian critique of force is perfectly valid against any other concept of explanatory intention. And in the second place, physicists continue e to speak carelessly about forces despite of Einstein and his imitators. Raymond A. Serway still enumerates as fundamental —besides gravity — the electromagnetic force between charges, the strong nuclear force between subatomic particles and the weak nuclear force (1985, 72). In general, atomic physics draws explanation with forces as if nothing had ever happened: e.g. Rittenhouse (EB 14, 330,2), Ziman (EB 14,340,1), Cuninghame (EB 14,361,1). And one speaks even about the "London Forces" which are applied between the molecules (EB 23,690 ,2) and the Van der Waals force is appealed to explain the liquids and solids formation (EB 14,377,1). The four forces that Serway enumerates appear to continue unperturbed on their explicative task: cf. Rosenfeld (EB 28,252) and Brown (EB 25,819).

One does not have the slightest idea when people like Rosenblueth say nowadays that science does not pretend to *explain* the phenomena. The concept of strength can only have the aim of explaining and nothing more. At least the discipline to which Newton dedicated his life never pretended anything else. On the first page of his principal work Newton identifies physics with the force science: "the rational mechanics will be the science of movements that shall result out from any sort of strength, and from the strengths required to produce any sort of movement" (1977, 1).

Newton was not so much impressed by the attack that Leibnitz formulates against him on its fifth letter to Samuel Clarke, according to which those forces are not in any way different from the scholastic's 'hidden qualities', by means of which it is tremendously easy to explain any phenomena: e.g. if a liquid tastes acid, then that can be explained by the hidden quality called acidity. That is the way Newton's explanations are: if the apple falls down to the ground, that can be explained by a certain attraction force whose single definition is to be the movement's explanation that is intended to be explained. It is like ancient medicine: if an ill person suffers melancholy, then that can be explained by the melancholic fluids.

Hegel says that it was Newton who "contributed the most" (GP III 231) to the inclusion of those brilliant explanations to Physics. But apparently, this way of thinking is very widespread: one can explain that molecules do not disperse themselves forming gases by recurring to Van der Waals' forces that keep the molecules together, which by the way is the explanation to describe liquidity. However, one sees that the only definition of these forces ends up being the explanation of the phenomenon which we wanted to explain. Nothing is again defined.

> With regard to the content, one does not say to us anything which the phenomenon itself did not have, that means to say, the phenomenon of the reciprocal reaction of these bodies in movement, only comes about in the form a determination reflected on itself, the form of force. If we then ask what kind of force is the force of attraction, the answer we are given is that it is the force that makes the Earth move around the Sun; that means, such explanation has exactly the same content which it had actually to explain first [...]. In ordinary life, these etiologies, which are the privilege of sciences, cannot remain unnoticed: they reveal themselves as empty chitchat and tautologies. If we ask why does this man go to the city and one responds us that there is a force that pulls that man there, we immediately see the absurdity of the answer, an absurdity which is not allowed in sciences (WL II 79).

In the same line that the Leibnizian critique, Hegel says that those qualities are not hidden at all: they are perfectly clear, since 'they do not have any other content than the phenomenon itself' which they intended to explain. Therefore they explain absolutely nothing.

"When someone inquires for an explanation, asks actually for the explicative factor a different content from the object whose explanation is being asked" (WL II 83).

However, "in the category of force there is not actually a concrete content" (VG 114); that is why "it has no other content than the phenomenon itself" (WL II 79), "it stems only from the same phenomenon's elements" (GP II 388). From this it follows that "the explanation by means of force is identical, formal, and with regard to the content is nothing" (NH 156).

The explanation is very generous: it postulates itself as explicative of an entity that cannot be seen, since that what can be seen is the phenomenon that is intended to be explained, and when someone asks for the definition of that entity in order to know what it is, then he is given the answer that nobody knows it. So, how can we understand what they are talking about if either sensibility or reason does no tell us? And what is the explicative capacity from a verbal maneuver that advances as an explanation an entity from which no one knows what it consist of?

This is why Hegel holds that, in balance, the only content that presents itself is the content from the phenomenon to be explained, since physicists are not capable of giving content to the entity called strength or force. With irony, Hegel calls these entities "determinations of reflection": not only because they are breed by the reflection of the abstract intellect that goes beyond what is observable, but because they are in themselves mere reflection or reflex of the phenomenon to be explained; they have exactly the same content; they are the "tranquil mimicry of the existing world of phenomena" (WL II 127).Whoever takes as valid such explanations, "wants to see *doubled* the *same* determination that is the content"(WL II 78).

It is not very flattering that the human mind has ever taken as a valid explanation any supposed entity whose proposers do not know how to define: "ordinarily, one says that we do not know the nature of force" (NH 156). One openly acknowledges that what we perceive are the effects of the force, that is to say, the phenomenon that is intended to be explained, but not the strength itself; this one remains as something unknown.

> It is frequent to say that the nature of force itself is unknown to us and can only be known in its manifestation. [...]. The content's determination of force is precisely the same as that of the manifestation. Consequently, the explanation of a phenomenon by means of strength is a hollow tautology (EPW. 136 A).

The really incredible circularity of this entire explanatory maneuver reveals itself in the first law of Newton called the principle of inertia. On the one hand, one defines an inertial system as that in which no forces intervene. On the other hand, force is defined as something that does not intervene in an inertial system. It speaks badly about the human mind that it formerly allowed these kinds of explanations.

And this is how Taylor and Wheeler keep speaking:

> To understand the nature of the concept, 'force', try to imagine how one could get along without it! Force is most obviously needed to explain why

a particle speeds up or slows down. A test particle, subject to no forces, is defined precisely by the fact that it does not speed up or slow down (1966, 101).

Let us clarify that. Strength is defined as something that explains that a particle modifies its speed. But a 'particle whose speed is not modified' is defined as a particle that is not subjected to forces. The emptiness and circularity of such lucubrations is manifest.

Of course, the attempt to define force as the product from the mass by the acceleration is equally circular and null, because it has been already stated that mass is defined as the resistance that an object opposes to the application of a certain force. Look how the *Encyclopedia Britannica* defines mass: "...the resistance that a body of matter offers to a change in its speed or position upon the application of a force. The greater the mass of a body, the smaller the change produced by an applied force." (EB 7, 915, 2). In fact, the mass is the quantification of the inertia, the quantitive measure of inertia, so that the definitory circularity between mass and strength is exactly the same that the one indicated between inertia and force: inertia is defined as a state in which no forces intervene, and force is defined as something that does not intervene in inertial states.

It is interesting to notice that the first law of Newton —the whole foundation of Physics- is one of the most metaphysic constructions ever formulated. It speaks about resting objects, but there has never been a resting object, since we know that lamp located over my table is travelling along with our entire planet and with ourselves at a speed of thirty km per second. It speaks about rectilinear movement but every movement we know are elliptic or parabolic orbits, and if something like a rectilinear movement existed, we would have no way of proving that it is indeed so, since the ruler or stick with which we would compare it is not evidently straight, and in order to verify that it is straight we would use another stick whose straightness is also unknown to us, and so on *in indefinitum*. Newton's defenders would say that we are dealing with relative rest and relative movement, e.g. everything is relative to the objects of the surroundings; but with such a statement they definitely ruin the law, because if there are any objects in our surroundings, they exert the attraction or repulsion strengths upon an object about which the law refers, and this law cannot speak about an object upon which no forces are exerted.

In fact, Newton thought in absolute movement and in absolute rest, assuming the myth of absolute space. But we already said (II, 1) with Hegel that it is not possible to speak about rest or movement, "since the movement is necessarily [directed] towards some place" (GP II 193). And that was in fact one of Einstein's first considerations: "Does it make any sense to speak about a movement which we consider to be unique in the universe? No, because one considers that an object is in motion when it changes position with respect to another one" (1984, 182). We must say the same about rest: An object is considered to be at rest when it does not change position with respect to other objects. That is why Hegel says that Newton's first law is "an empty mental monster" (GP II, 193).

That first law tells us that there is a pair of phenomena that need no explanation: rest and uniform rectilinear movement. But these two phenomena do not exist! It can only be brilliant that they tell us that no explanation is needed for everything that does not exist!

The circularity and the absolute uselessness of the first law transforms the pseudo-concept of force into something absolutely circular and uselessness, because only that law could give any meaning to the word force.

Undoubtedly, the general omission of a genuine definition of strength has been present: it is the delusion that we have already denounced many times: one believes that a meaning can be provided to "force" by only pointing out to some empirical data. In the combat to this deceit, Einstein's contribution has proved to be particularly valuable. It is true, however, that the decisive and scientific warning had already been stated by Hegel a century before: "Force [...] is not something perceptible; therefore, empiricists surrender themselves to such determinations in an uncritical and unconscious way" (GP III, 84). But Einstein managed to reach the wide public, or at least, the public who fears Philosophy but not Physics. Einstein´s argumentation proves one out of two issues which for our practical effects are equivalent: either that force is not empirical data, or that the empirical data that some call force does not correspond in the physical world to any reality

It would have been enough to indicate that the experience of effort is introspective; therefore, it is not empirical, since one understands by empirical what one perceives by means by some of the external five senses.

Besides, it would also have been enough to indicate that one understands by 'force' the 'capacity' of producing certain effects, that means to say, the power of causing certain manifestations. But we have already stated (III, 9) that a *power* is not an empirical data. Capacity means that something *can* produce certain effects, but sensibility witnesses, at best, that things *are*, but not that they *can*. If we want, we will *infer* that they can, but this demonstrates that a power is not an empirical data, for it needs to be inferred. Therefore, the only capacity and power that we actually perceive is the self-determination of spirit: its real capacity of producing determinations that did not previously exist. We will get back to this point. We will return to this point, because the causality concept had imposed the same result to us: Whoever wants to explain what the 'becoming' of something new in the world is, cannot do without the only cause that makes exists what did not exist before. It actually does not do without anything: the proof thereof is that the concept of force is explanatory to us —a concept that stems not from empirical data but from self-consciousness.

The first of these two observations —which would have been enough to prove the others wrong— had already been made by the most reasonable physicists. Max Born affirms this: "The fundamental concept of statics is force. It is derived from the subjective feeling of exertion experienced when we perform work with our bodies" (1962, 14). And Eddington completes:

> The primary conception of force is associated with the muscular sensation felt when we make an effort to cause or prevent the motion of matter. Similar effects on the motion of matter can be caused by non-living agency, and these also are regarded as due to forces (1978, 57; orig. 1920).

Neither Born nor Eddington seems to ponder on the fact that attributing effort to material things is animism and primitive mentality. Now, the only content that these two authors are indicating as the meaning of force is precisely effort. This is the reason why the Leibnizian critic of Newtonian strengths mentioned the 'hidden qualities'.

One must have in mind that when Margenau affirms that "Nobody has seen this force of physical attraction, and no one ever will." (1978, 335), he is properly saying something that any reflexive person could deny: by definition, what is seen are the effects of force, while force itself cannot be seen. And what is said about seeing must be said about touching too: the tactile sensation itself is the *effect* from the force one assumes to exist; force is conceived as the capacity to produce these effects, and the capacity cannot be touched. No physical force is empirical data. It is a mental construction that we imaginarily project on the physical world, for which it is required that the content of the projected concept is extracted from somewhere else.

Inferring that physical strength actually exists is another thing. Following Einstein, however, we must warn everybody that such inference is illegitimate. The syllogism should contain a universal affirmative premise, easily refutable by a single negative case. The premise would have to say: 'Every time there is impression of force, force exists'. In order to refute this, one only needs to bring up the experience that everybody has had in the train station while waiting for our train to start moving: if the train that is next to us starts to move and we see that through the window, we have the impression that we are moved by a force of acceleration which actually does not exist. A single case like this is enough to prove that perception about a physical force is a subjective impression provoked by the peculiar manner in which are our anatomic organs constituted and to their physiology.

Curiously enough, the experiments prove that the sensation we have of the force of gravity is highly unreliable. The experimental psychologists and physiologists have achieved to tackle a man just by making him see things. With his two feet well set on the ground, this man is shown by means of a cinematograph what our eyes see when the car in which we drive stops violently or makes an abrupt turn. This man *feels* the force that impulses towards the right, and his muscles, in order to keep the equilibrium, apply all their effort to the left, and when that happens our man falls down to that side. The force that he felt did not exist. Again: there is the sensation of force, but there are no forces. The alleged premise is simply false.

One should take into account that in the sensation of angular, rotational or gravitational *momentum*, the correspondent organs are certain ear parts called bony labyrinth and vestibule, with expansions called the saccule and the utricule. The labyrinth is composed by three semicircular canals whose planes lie approximately perpendicular to each other. All this interconnected forms a cavity that is filled with a liquid called endolymph and the cavity is covered with hair cells that have nerve terminals on their base which transmit electrical signals to the brain in function of the movement of the endolymph, a phenomenon that occurs when the body accelerates or stops completely. As a matter of fact, perceiving static pressure or gravity is a task of the the saccule and the utricle, whose hair cells are covered with a jelly like liquid in which little calcium granules float; these granules apply pressure to the hair cells. But we must state three very important things about this organ:

1) The information transmitted by this entire organ is proprioceptive, not exteroceptive like the one that is transmitted from the Cochlea, since the eardrum perceives the vibrations from the surrounding sonorous environment, not the vibrations of vellum hair within the organism.

2) The entire system is connected with the reflex centre which governs the eyes, the neck and the extremities. This is what explains the experiment of tackling down a man by visual data.

3) There are several tricks in order to make this organ work without the existence of the forces that he forces himself to believe that exist, e.g. with electrical energy, with chemical substances, drugs, and even with hot and cold water, since temperature changes make the endolymph circulate and put more pressure to the granules.

The thesis according to which there are forces in the physical world would have to be based on other reasons. To affirm that they exist because 'we feel them' is not a good argument, because, obviously, we feel them sometimes despite they do not exist. Such an argument would be as unscientific as saying that there are surfaces because we feel them; in physics this argument in favor of surfaces has never been acceptable.

A different argument would be to pledge that the existence of forces in the physical world is required to explain the phenomena. But Hegel attacks this brutally as we have seen. Who can accept as explanatory an entity nobody knows what does it consist about? It would be useless, of course, an exit like the following one: We know that there is a force, even though we do not know what it is like. That is tantamount to say: We know that blictiri exist but we do not know what blictiri is. It is obvious that there is no trace of knowledge in any of these intellectual whims. It is impossible to know if there is or not an X, when that X is not given any content.

And if no content is given to the word 'force', the only content that is present is that of the phenomenon we wanted to explain. And this is a mirror game: they provide us as explanation the same fact of which we demanded an explanation: "properly speaking, there is no concrete content in the category of force" (VG 114). And if someone pretends to indicate any empirical data as a content we can reply what Hegel already said: "force [...] is not something perceptible" (GP III 84).

4. Explanatory factors

Now, the concept of force is paradigmatic. It would be a mistake to believe that Hegel's attack is invalid because physics have substituted the concept of force by the concept of *field, energy, mass, momentum, conservation, probability and law*.

In first place, as we demonstrated, atomic and molecular physics still employs force carelessly. And in the second place, that is what is penetrating of the Hegelian analysis: all concepts of reflection (*Reflexionsbestimmungen*), which are supposedly explanatory, have the same content, or more precisely speaking, the lack of content that the concept of force has.

One only needs to mention what says Hegel about the last concept we mentioned in our list, namely, the concept of law, which is by the way the decisive concept in physics as in the other natural sciences, so that his critic concerns to all of these disciplines: "*forces have exactly the same form as law*" (PG 119); "both have the *same* content" (*ibid*); "the explanation includes the *law* in the *force*, as in the essence of the law" (*ibid.*).

It is very important to understand what has just been told to us. Whoever advances a law as an explanation of a certain phenomenon does not evidently mean that a *mental* operation called judgment —for the law is a judgment with a subject, a verb and a predicate— suffices to produce a real phenomenon whose existence needs to be explained. And obviously it does not mean either that a certain set of words, which would be the expression of the law in mere sounds and ink– would suffice to produce the real phenomenon in question. On the contrary, what the person that advances a law as the explanation of an observed phenomenon wants to indicate is that something real is intertwined with the empirical facts and that, without being itself an empirical data, is capable of producing the empirical effects in which the phenomenon consists. That the law is not an empirical data has already been expressed (III, 9): the law contains an 'all' or an 'always' and it is impossible to observe those contents by empirical means. Nobody knows exactly what it signifies, and yet nobody doubts that it exists and produces empirical effects. And those are the same characteristics of force according to our study: a real entity that is behind of phenomena, that is not observable in itself; an entity of which no one knows what it consists of and yet has the capacity of producing phenomena and explaining them.

This is why the Hegelian critique against force is also valid against the law and other allegedly explanatory factors: as explanation of a certain phenomenon we are given an entity whose only definition is to be explanation of the phenomenon.

We pointed out that, in order to avoid such critique, it would be useless to pretend that the law is not explanatory but only descriptive. A proposition that says 'always' or 'all' cannot be descriptive. Not a single verifiable fact can be described with those words. Now, in order to scientifically and justifiably advance a thesis, this thesis has to be an object of experience or something that is necessary for the object of experience to exists, that means to say, a factor that explains the existence of the phenomenon; otherwise, what one affirms is nothing more than a whim, an irrationality that has nothing to do with the work of science.

This dilemma is indispensable to understand the systematization of concepts started by *The Science of Logic*. The book of being deals with the concepts which are supposedly immediate or descriptive: we have seen quality, quantity, continuity, discretion, intensive, extensive, limit, finite, and being itself. Hegel proves that the alleged immediacy is merely imaginary, that all are mediated and lack of meaning if the mind does not go the next step forward. The book of the essence deals with the concepts supposedly explanatory, mediated determinations or reflections which do not have their origin in observation. One demonstrates there too that they do not explain anything and that they lack meaning if they do not reach the spirit (the concept), which is the subject of the third book.

As a guide we could have these two sentences: "Truth is the essence of being" (WL II 3). "The truth of being and essence is the concept" (EPW 159)

It could seem that the object of essence is outdated since no science speaks nowadays of essences. But that is only a superficial consideration. The first thing that a scientist does, consciously or unconsciously, when he deals with a baroque set of empirical impressions, is to precise *what are those impressions about*, that is to say, he tries to determine what kind of object or fact lays before him. In other words, what he must do is to determine the essential. It is the Kantian problem of subsumtion. The essence is the universal concept under which we subsume the phenomenon: Is it a rock? Is it a tribunal? Is it a circus? Is it a book? Is it a joke? Is it a promise? Is it a requisite? Is it a whim? Is it a wedding? Is it a landscape? Is it a nightingale? Is it a speech? Everything depends on this essential discernment which, by the way, is closely related to the theory of Gestalt, but it is not reduced to the visual, since it comprises all things intellectual and human. We had already quoted this text: "It happens as always that one alludes or mentions a perception or an experience; so soon man has spoken there is a concept there, there is no way of making it aside, it reappears in the mind as a clear sign of universality and truth, since it is precisely the essential" (GP I 336). The word essence is maybe obsolete, but the concept of essence, it doesn't.

Now, "the essence is determined itself as explanation (*Grund*)" (WL II 63). This is why Hegel arranged the supposedly explanatory under the captaincy of essence. By definition, the essence of a body explains why the body is like it is. That would be tantamount to explain the most important thing, but evidently, that does not happen here.

When I say that what I have in front of me is a tree, I explain the multiple and varied empirical impressions that are coming to my senses. I do not only orient myself in the middle of a chaotic parade of sensations. Impressions are what they are *because* what I have before me is a tree.

Therefore, the affirmation according to which being and essence complement each other has two senses. First, this affirmation means that we cannot describe without explaining; the crucial thesis that entirely refutes those who believe science must be reduced to descriptions. In order to describe we need to use some universal concept, and since this concept tries to grab the essence, one infers that it must have an explanatory purpose. Second, the truth of being is the essence because (cf. Zubiri) the essence is in the being itself making it being what it is, so that other aspects or non-essential details are expendable and irrelevant: they are mere appearances.

But this is only a mirror game according to Hegel. Just as in the case of force and law, so the essence is presented to us as an entity that lies beyond the phenomena and explains them, but it is an entity that nobody is able to explain. It offers itself as the explanation of the phenomenon, but its only definition is to be the explanation of the phenomenon. This identity between essence and phenomenon has been commented at length, as if it were a thesis that Hegel defended as a part of his system; in reality, that very identity is what Hegel *criticizes* in the concept of essence in order to reject it.

Hegel rejects every allegedly explanatory concept because the only true explanation of the world is the spirit. We immediately see how this critique is valid against all the above mentioned concepts: field, energy, etcetera. But we need first to examine more closely an interpretative question. Commentators did not perceive that the book of essence is that of the allegedly explanatory concepts; they missed the fact that this book is divided in three parts: essence, phenomenon and reality. The union of the essence and the phenomenon is reality, as synthesis of the thesis and the antithesis, so that all this movement constitutes reality; it explains it in the mind of those who believe in the above mentioned explanatory concepts. In addition, everything leads us to the most explanatory concept we can think of, which is no other than that of cause, and whose critique we have made in our third chapter.

We already said why Hegel named that entire treatise under the name of essence. As the encyclopedic summary tells us, the essence "is, essentially, explanation (*Grund*)" (EPW 121 A). And the treatise itself tells us: "the essence determines itself as explanation (*Grund*)" (WL II 63). But a simple glance to the concepts there contained is enough to persuade us that they are only allegedly explanatory concepts. 'Form' and 'matter' since they were coined by Aristotle, have had no other purpose than explaining facts. The conditions in which the existence of a being depends, contribute also to the explanation of that existence. The properties of a being explain why such being operates as it does and how it is in fact constituted. The principles of identity and (no) contradiction evidently want to explain why things are this way and cannot be otherwise.

The category whole-parts: certain being is like it is *because* it is constituted by certain parts, or because it forms part of a certain whole. An attempt to explain its existence was to evaluate a being in terms of 'possible and 'necessary'. And let us not speak about categories like force, substance and *Grund*.

The substantive *Grund* and the verb *begründen* need to be succinctly clarified. The term 'explanation' has two different usages that depend on the nature of the grammatical subject (if it is a person or a thing).

For instance, one person explains certain fact by indicating its causes; but we also say that the cause explains the existence of the effect. The German verb *begründen* has also these two usages. The substantive *Grund*, in its material meaning, means a ground, a fundament, or a basis. In theory, *begründen* would mean to lay the foundations of something, but in its scientific meaning *Grund* is a real entity by means of which one explains the existence of an entity or an event. In regard to that meaning, one would have to translate *Grund* as 'explanatory factor', but we also say that the cause as real entity is the 'explanation' of the effect. This is how we have translated Grund in the previous pages, because 'laying the foundations' would be only a metaphorical language that is not pertinent to the question. That we are loyal to Hegel's thought is confirmed by the titles of the two annotations of the section called the "The Determined *Grund*"; in both of which the word *Erklärung* —which means explanation— appears. Besides, this is corroborated by the fact that all the systematization effectuated by The Science of Logic would lack if the second book did not thematize the explanatory concepts, in contrast to the first book which thematizes the allegedly descriptive ones. Without this dilemma the whole work lacks its sting of truth.

But let us deal now with the modern explanatory concepts.

We lay aside the *imaginative* figurations, which can be varied. The *figurative imagination*, of field definition, in physics has the same content —or lack of content— that the definition of force: an entity that is not empirical in itself, that means to say, that is beyond or underneath the physical manifestations, and whose only characteristic is the capacity of producing this empirically perceptible manifestations. What we *see* are the metal arrows that are orientated towards the poles, but in moment at all do we see the field itself. One even calls it a 'field of force'; one conceives it as an aggregate of forces. D'Abro says something very illustrative in regard to gravitational fields:

> For instance, the force of attraction exerted by the sun on a planet varies with the position of the planet. Let us observe that the gravitational force is disclosed only through the behaviour of the planet; nevertheless, we may reason as though a force were still in existence at each point of space around the sun even in the absence of any planet. We are thus led to conceive of a region of space at each point of which a force is present. The aggregate of such forces is called a *field of force*. (1952, 215). (The italics are mine)

To be sure, in electrostatics and electromagnetism one deals not with masses or planets but with charges.

The empirical data is the trajectory and movement of the planet. The field as such is not an empirical data; it is an entity that is posed in order to explain why empirical data are as they are. In this regard, D'Abro makes also an interesting point:

> The magnitude of the force at the various points of a given field is proportional to the mass of the particle on which the force is acting. There is no sense, therefore, in attributing any definite magnitude to the force at a given point until the mass of the particle has been specified. (*ibid.*).

It turns out that the same force of the field is something indefinite; it is not an entity with determined characteristics; it depends upon the 'patient'; the latter determines the magnitude of the form. The mirror game which we have previously criticized could not be more obvious: the allegedly explanatory entity does not have any other content than the phenomenon it intended to explain; we only have the illusion that such phenomenon has been explained, because we have added an entity whose only content is to explain it!

As a matter of fact, to speak of a field in this regard is to employ and metaphor and to start poetizing. In its proper sense, a field is a plot of land, an inhabited portion of an earthly surface. What physics mean to say is a certain set of forces, and I do not manage to conceive how they could possibly come to believe that they have abandoned the 'obsolete' Newtonian concept of force. The difference between force and field consists in that physics believe that they *imagine* the latter as occupying a region of space, while force seems not to have extension. But since the field as such is not visible, then it is not imaginable. The same thing happens as with space: there are people who believe to see the space that exists between that table and the wall, but the only thing he sees is the table, the floor and the ground. The Newtonian scientist affirms the action at distance evidently holds that a force *is extended* through space from one body to another so that, in terms of extension, there is no difference whatsoever between them.

Margenau's next quote converges with what D'Abro said:

> A non-material field is not descriptive of any material property (although it is usually caused by matter) but describes some latent effect that would

> take place at a point of space under certain circumstances. [...] Nothing, however, may actually exist or be evident at that point. What matters is that *if* a charge were placed at that point defined by x, y, z it *would* experience a force proportional to the function of the field. (EB 25, 823, 1)

From the above we can say that the content of field poses unrealistic conditionals. D'Abro only ascribed fields a hypothetical 'as if' content.

On the one hand, the field does not exist where (space) and while (time) it is not exerted. On the other hand, in the moment in which it is exerted what we have there is a force and only a force. But the Newtonians that speak of action at distance also hold that the only thing that exists between bodies is force. I do not see in what sense does the physics of fields has abandoned the concept of force, nor in what sense it has abandoned the concept of action at distance. Besides, we have shown (III, 8) that this entire problem in regard of the action at distance renders itself as a pseudo-problem when we realize that contact does not exist and that the difficulty lies in action as such, not in action at distance.

We do not need to go back to the concept of mass —whose defenders regard it as a truly explanatory concept—, for we have pointed out that force comes within its very own definition. Since the definition of force lacks content, the concept of mass is also undefined. By the way, it is noteworthy that mass is directly conceived as a resistance force that is opposed to displacement, and that inertia itself, whose quantification is mass, is also commonly conceived as a force that resists the change from rest or movement. The only thing that physics perceive in all this business is the displacement and the duration of it: mass as such is not perceptible. It is an entity that is posited in order to explain the empirically perceivable data.

Since *momentum* is defined as the product of mass by acceleration, the moment implies all the force that is implicated in the mass. Therefore, it carries all the lack of content and all the tautological game of mirrors that the explanatory entity called force entails.

Since Einstein demonstrated that energy equals mass multiplied by the speed of light squared, we have to say the same in regard to energy itself. All the lack of content which characterizes force enters in the definition with mass.

But even leaving that aside, it is perfectly obvious that energy is identified with force, for it is commonly defined as the energy of carrying out work. Work is the effect, the observable phenomenon, the translation of a body from a position to another that is even higher. In addition, we saw that force is commonly conceived as the capacity of producing certain observable effects: the 'power' of causing certain manifestation. It follows that energy, in perfect identity with that Newtonian force that physics naively believe to have abandoned, is a non-perceptible entity whose existence is posed in order *to explain* the empirically observable phenomena. This is why Hegel warned us that the concept of force "is the most prominent one" (GP III 84), stressing the fact that it is not properly a concept but rather certain 'way of thinking'. What physics has made after Hegel is to shuffle in very different ways this same concept or lack of concept, in the hope of inventing new and different ideas.

By the way, what the popular formula of Einstein affirms is that mass transforms energy and vice versa: as statement that seemed to be fabulous in its times and which is still regarded so, but if physics with true scientific rigor demanded themselves true definitions with contents, that would not need to surprise anybody, for both mass and energy are always force: a posited explanatory entity that lacks content.

The only difference that exists consists in the diverse kind of measurements that scientists whimsically decide to carry out. However, one should notice that they are not measuring mass or energy themselves, in spite of the fact that they believe to be measuring that. In the case of mass, what they measure is the acceleration that a body suffers when certain force is applied to it, that means to say, they measure the space covered by the body in a given time; in other words, they measure a length and they divide it in a certain number of seconds. Best case scenario, what would be empirically measurable is length and that thing which physics call time: mass in itself is not perceptible or measurable. In the case of energy what they are measuring is work, that means to say, the distance along which a mass has been moved and the duration of that movement; the bigger the distance the bigger the work; the bigger the time the smaller the work; length and time, that is what they measure; energy itself is not perceptible or measurable. It all depends on the imaginative combinations that physics make with length and time; they call one certain combination mass and the other energy; but that one combination transforms into another should not surprise anybody, since they are all mere multiplications and divisions that one can make with a pencil and a sheet of paper. Since they are only force, mass and energy themselves lack content; and no one should be surprised by the fact that a thing that lacks content is transformed into a thing that also lacks content.

Since 1860, and specially since 1890, the law of conservation of energy has become the touchstone of physics and perhaps of all natural sciences up to the extent that Max von Laue says that one intends to deduce from it the rest of all natural laws and constructed an entire worldview around its form (cfr. Schilpp II 1970, 515). As for physics in particular goes, whoever studies these treatises corroborates that they are constructed upon Hamiltonian equations, and all what these supposes is that the sum of kinetic energy and potential energy is unalterably conserved.

One can appreciate in Taylor and Wheeler to what extent the enthusiasm for the law of conservation of energy has escalated. Defined in Newton's terms as the product of mass by speed, the *momentum* is *not* unalterably conserved in the collision of particles that travel near the speed of light. "We must therefore choose: We must abandon *either* the Newtonian expression for momentum *or* the law of conservation of momentum. The law of conservation of momentum has become so important to us that we shift to it as a new foundation. We *start* with the law of conservation of momentum and from it *derive* the expression for *momentum defined as that vector quantity which is conserved in all frames of reference.*" (1966, 102)

In other words, *momentum* is defined as what is 'conserved'.

At first sight, the person that comes to us, boasting about his large scientific experiments, seems to possess a very profound knowledge of the universe, especially when he says to us that the existing quantity of energy in the world does not decrease or increase: it is always conserved in his opinion. His words are sonorous and impressive; but if we ask him what energy is, what that august, impressive and unalterable thing is, he answers to us: what is conserved.

The law of conservation of energy has this grandiose content: one conserves what one conserves.

To make such statement one does not need to carry out the most elementary experiment; one does not need either to have any knowledge of reality whatsoever or to open his eyes to look at the world. One only needs certain intellectual masochism to take pleasure in tautologies.

The sharply tautological character of the law of conservation of energy was obvious before Taylor and Wheeler made their mortifying choice. If the term energy does not have content and yet one affirms that 'it' is conserved, the only thing we are told is that the 'it' is being conserved. And the proposition is summarized thus: one conserves what one conserves.

Since the first time it was formulated, such thesis was an *a priori* proposition which was by definition unverifiable. And if it refers to the quantity of energy in the entire universe, one would have to measure the quantity of energy in the entire universe: a task that cannot be completed even by all humans. Even worse: they would have to measure it one minute later to see if it has not increased or diminished, something which would require *another* team of humans just as numerous because the first team would has not yet finished its task. The law is also unverifiable by definition in the case that it refers to an isolated system or a tiny region in the universe. Even if we were to suppose that we would carry out a measurement in this instant and another one after ten minutes, the thesis would not be probed thereby, because the quantity of energy could augment in the meantime and return to its previous quantity in ten minutes. The verification would suppose the paroxysm of a measurement indefinitely repeated, which is something impossible not only for technical reason but by principle: the *proccessus in indefinitum* cannot be completed. Let alone the problem of defining what physics call an isolated system, because everything seems to indicate that they define it as a 'portion of the universe in which the energy does not increase or decrease', which would render us this wonderful definition: in a portion of the universe in which the energy does not increase or decrease, the energy does not increase or decrease. Indeed, they only know that a system is isolated because of the fact that the energy does not increase or decrease in it. As for the imaginary or real isolating surfaces that limit the system goes, physics only know that they are isolating it because in its interior the energy does not increase or decrease. Therefore, the insolently and sensational tautological formulation we just mentioned is unavoidable.

In the entire business of the conservation of energy we find tautologies everywhere. If the particles —which according to the law should be created at a given instant because the observable particles have lost their energy— are not observable, the physics hold that the new ones commence to exist 'virtually', not effectively nor observably; and then, of course, the sum of all the energies in the system remains unaltered. In order to keep the equality unaltered we can always imagine that the electron is surrounded by a cloud of 'virtual' photons: that is what

Rosenfeld calls the "dressed electron" (EB 28, 250, 1-2). If we do not need this virtual energy to equalize the sum, then we will simply consider that the electron is 'undressed'.

If the number of things that I can call energy is unlimited, I do not have any problems whatsoever to make my results match. But, naturally, if they do not have any common denominator (and they do not have it, because they have not defined energy), if the unlikeness and heterogeneity between them is unlimited, what is conserved is an abstraction; the energy is perhaps the most abstract abstraction ever invented. In fact what is conserved is the capacity of the intellect for making abstractions; what is conserved is the spirit.

We have gone through the principal explanatory concepts that have come into vogue after the death of Hegel. In all of them one confirms the Hegelian demonstration of the tautological failure of all scientific explanations. It is impressive to see how quantum physics have arrived to the conclusion that the physical explanations have failed. It is not a change of paradigm, as Kuhn would want; it is something much more earnest, so earnest that Einstein died without accepting it, despite all the efforts of Bohr to convince him.

Classical physics 'explained' fundamentally by saying that there is nothing to explain: it explained by continuity between a past state and a present state of a physical system; quantum physics breaks with this continuity and hence it tears away the possibility of explanation. Margenau says correctly: "If there were gaps in this understanding, missing links in the chain of continuous action, the term causal would not be applied to it." (1978, 175) What quantum physics discover is that there are leaks; they find that some links are missing. With regard to the luminous phenomena and their dependency on experimental devices, Bohr himself states: "this real situation obligates us to renounce to a rigorous causal explanation" (1964, 8).

Commenting Bohr, Weizsäcker says:

> ...we are forced, not to renounce classical models, but to renounce models. Nothing like a quantum-mechanical model which replaces classical models and which then would admit of description of nature in terms of an explanation by the quantum-mechanical model, exists. (Bastin 1971, 326)

Now, if the physical explanation of the world has failed —and Hegel demonstrated that it had to fail— then the only explanation of the world is the spirit, or, as it is expressed in the *Science of Logic*, the true of the essence is the concept (= spirit). Zubiri did not understand this.

The concept 'force' —which, as we saw, is the most eminent of the explanatory concepts— is in short a projection of the concept, which is known directly in the self-consciousness of the cause that determines itself, namely, the self-determination of the spirit. "Will is power in itself, and it is the essence of all power, both in nature and in spirit" (VG 113). "The subject is what is meaningful to itself and what is explained by itself" (Ästh I 435). Reason identified with method is the "*supreme* force or, more precisely, the *only* and absolute *force* [...] (WL II 486).

Whoever thinks he/she can explain differently the production of something entirely new in the world is employing a concept of cause (III, 8) which cannot be given any meaning. How childish is the process by which some think a phenomenon is determined by another phenomenon —in a magical transmigration of properties— indefinitely, without ever reaching a being that determines itself, and for that reason the entire set lacks determination and remains unexplained. The only source is the spirit: the being that determines itself.

The difficulty that some people bear to accept the Hegelian thesis is a problem of imagination only. We would like to say stress the merely imaginative character of this difficulty which is in itself a triviality but which is very widespread. They suppose that the world is "outside" from the spirit, and they do not realize that this expression *lacks meaning completely*, for the spirit is not a spatial thing of which one can speak of an inside and an outside. The objectors of Hegel are imaginatively creating distances and distinctions that do not even exist.

It is inspiring to see that a physic like Henry Margenau has understood that difficulties of such kind lack all kind of meaning:

"As the majority of scientists, Einstein did not solve the basic metaphysical problem that underlies all science, namely, the meaning of exteriority" (1978, 249).

If, in contrast to idealism, realism consists in saying that physical world is 'outside' from the spirit, then it is a thesis which does not have any meaning whatsoever.

Public opinion was shocked when the quantum physics experiments revealed that the electron becomes a bodkin only *because* the subject chooses to observe its position and that even, as Heisenberg said, "its size depends on the experiment that we carry out" (1930, 34). And this does not go only for quantum physics. As Eddington pointed out "...the relativity view is that a field of force can, like length and duration, be nothing but a link between nature and the observer." (1978, 43; orig. 1920) Max Born also said that "A gravitational field [...] has no meaning at all independent of the choice of coordinates" (1962, 345). But the universal astonishment—even among the physics themselves—evidently had as its cause the so-called belief according to which the world is 'outside' from the spirit, which is pure and sheer non-sense. And if Einstein himself was scared because he thought this was 'telepathy', then we can only conjecture that he was imagining that the physical remained far away from the spirit.

It has been a tremendous mistake to believe that Hegel denies the reality of the physical world. What he denies is that 'being real' means 'being outside'. It is the spirit what makes real the material, which means that the material is *real*. "Natural things are false existences; that does not mean they do not exist, but rather that they do not have their truth in themselves" (EGP 116).

5. Probability

Probability deserves a whole different treatment. According to some, it is a theory that is explanatory. Besides, it is a theory on which biology has a keen interest, especially in regard to evolution, which is our next subject.

A probabilistic law is imaginary projected as a real factor which is not empirical in itself but which 'explains the empirical' data. In a like manner as with determinism, one supposes that probability is a real entity that works among and *in* the things themselves, and that it causes some effects which are the phenomenon, which in this case is a certain frequency of events. Of course, the mirror game and the 'double seeing' is just as true as the other allegedly explanatory entities which we have considered, because the probabilistic law has the same phenomenon it aims to describe. In this point, there is no difference between a probabilistic and a necessary law, because one supposes that the observed frequencies necessarily follow from the 'objective probability': that is the myth of what is 'unpredictable but unavoidable' of Manfred Eigen. In the same line, Mario Bunge says the following: "In short, our version of QM is as deterministic as classical mechanics [...] as soon as the probabilities are both objective and lawful, indeterminism evaporates and stochastic determinism remains." (1973, 100) Such an illusion should not cause any wonder, since the dilettante-philosopher speaks of the 'law of the big numbers'.

Before analyzing the concept of probability, one should notice that the *explanation* by a probable law demands from the mind a bigger desire of self-deceit as the explanation by a necessary law. If an 'it always happens like that' does not explain nothing, an "it frequently happens like that" explains even less. We said that if we ask why it thunders when it rains and one answers to us: 'because every time there are black clouds it thunders' we have not received any explanation; but if one answers to us 'because sixty percent of the times there are black clouds it thunders', the explanatory nullity would be even more accentuated. In that case, we could even speak of an anti-explanation, because what one should really be explaining is why sometimes it thunders and why sometimes it doesn't.

Statistics is a technique, not knowledge. It is an effective way or proceeding in the practice, but it is not an explanation of reality. By means of a statistical law, we could *foresee* how much percentage of the population studied will adopt certain conduct, but by any means we could *explain* why it adopts it. As we have said, the probabilistic thinkers *figure* that there is a being or a real factor (the 'objective probability') among things whose influx explains why seventy percent of the times something happens and why thirty percent of the times it doesn't. Even though we supposed such entity exists, it is by no means sufficient to explain the observed frequencies. If the *entire* population is under its influx, but some individual behave in the observed way and some of them don't, then it is not its influx what makes that, since in theory both of them are under its influx. As Hegel says "amounts to no more that the *great influence* of environment; and this does not tell us what does and what does not strictly belong to this influence." (PG 194). Such entity is not even explanatory for the case of a majority: the only cause that would be explanatory is that which tells why A proceeds in the observed way, and B do not behave in the observed way even though A and B are under its same influx. The influx of such entity does not suffice to explain the behavior of A, because B was also under such influx and did not behave like that.

Let us go to the concept of the probable. Hegel did not treat it, but he mocked the concept of the 'possible', making clear thereby that it is a replica and a poor copy of the existing; it is merely an speculation of the real, which is projected by philosophers as if it was an entity in the world of the 'possible'. The possible has the same content than the real, but it is projected to an inexistent world. Now, the multitude of the 'possible' is divided into two groups: the probable and the improbable. For something to be probable it needs to be possible. If the possible only has meaning as a product or as a reflection of the intelligence, the probability deserves the same luck.

Possible is that which 'can' be. But we have stated that a 'can' does not have an empirical meaning, and also that in order to infer an 'is' from a 'can be' we need to take the content of can from someplace else, for the content and the meaning are not in the empirical data (cfr. III, 9). Now, if the concept of 'can' is not of empirical precedence, its origin is the reflection of the subject upon himself, and its meaning is that which we know by self-consciousness: the very real *power* that has the spirit of producing determinations and experiences that were not there before, that is to say, the causality of the spirit. That something is possible means that the spirit can do it. The word 'possible' lacks meaning completely if we do without the spirit.

That something passes from being inexistent to be real is a fact that evidently requires a cause, for that which does not exist cannot do anything. Now, to say that something is possible is tantamount to say it can go from being inexistent to being real. The real possibility is not reduced by any means to non-contradiction. It is obvious that the capacity of existing —which is characteristic of the possible— does not lie in it because it does not exist, but rather in the cause that *can* make it being.

Therefore, to speak of probabilities and possibilities *before* the spirit existed is tantamount to utter non-sense. If there is something that cannot be the product of evolution that thing is the spirit.

Fortunately, Niels Bohr the most intelligent physic of our century, and his Danish school understood with perfect clearness right from the start of quantum physics that "probabilities concern only mental states; a probability value can only measure the strength of our belief and the accuracy of our information" (Bunge, 1973, 66). We said that the word 'possible' lacks meaning if one does away with the spirit; the same thing goes for the probable, for that is only a species of the 'possible'. By definition, what is probable does not exist yet. Consequently, one cannot say it exists independently from the intelligence that considers it probable.

In order to determine the degree of probability of something, all the relevant information has to be taken into account. To judge whether a piece of information is relevant or not is a prudential consideration that inanimate things cannot obviously make. The intervention of intelligence is necessary.

But there is more to it. I can judge whether a piece of information that I possess is relevant or not, for I do not have it before my mind; therefore it is impossible to gain absolute and mechanical certainty as whether or not we have obtained all the relevant information. For pragmatic needs, because I cannot remain indecisive all my life, a moment comes in which I decide prudentially that the information that I have obtained is sufficient to me. In function of this, I calculate the probability in question and proceed. Despite what the 'objectivists' may say, the only probabilities that exist are the ones which are built up the way we just indicated. Many operations of the intellect intervene in their constitution.

To think that there is an entity called percentual probability within things is to fall into the illusion denounced by Hegel; one projects imaginatively an entity which is not seen but which is 'under' the phenomena that we can see, and whose only definition is being cause or explanation of these same phenomena.

In order to make probability something 'objective', the following thinkers want it to be a propensity or a tendency: Smoluchowski, Poincaré, Popper, Margenau and Bunge. But none of these authors has paid attention to the fact that a propensity or a tendency is not empirical data; not even the tendency to eat called hunger is an empirical data. The meaning of these words is something known by reflection of the subject towards himself. Here we are before the inwardness and the subjectivity that objectivism pretended to dismiss. Naturally, it would be very equivocal to attribute propensities to what is probable in itself, for that which does not exist yet cannot have any propensity. One should attribute them to the causes, but if these are material and physical objects, such attribution would be animism, and H. R. Post has mocked ""... the pagan device of investing the world of phenomena with pervasive wood spirits called propensities." (Bastin, 1971, 279). The causes would have to be *true* spirits, and there we find again the subject which was trying to be avoided by the objectivists.

The above mentioned authors do not realize that the propensity which they affirm is a propensity towards existence, and hence it would be a propensity of a fact that does not yet exist; it follows from this that neither the fact nor its property can be the real referent of the word probability; the only real referent that exists in this moment is knowledge; therefore, probability can only mean 'the strength of our belief and the accuracy of our information', that is what the Danish sustain.

On the other hand, the *Spielraumtheorie*, the range theory, was held with variants by Bernoulli, von Kries, Bolzano, Waismann, Wittgenstein, Keynes and Carnap. Although its ambition went far beyond, this theory only tells what the expression 'percentage' means, which is the expression of the degree of probability. This theory tells us that the percentage is the quotient or fraction whose numerator is the number of the favorable events, and whose denominator is the added number of favorable and unfavorable events. In order to talk about probability, this theory insists in that the number of unfavorable events must have *a priori* the same probability of occurring than the favorable events. It is the famous indifference or equipossibility they are always speaking about.

This has been acutely observed, but as von Wright notices "The question may be raised whether randomness and equipossibility can be satisfactorily accounted for without reference to states of knowledge or ignorance." (EB 23, 631, 1s.) It is evident to me that the amount of events covered by the so-called denominator cannot be determined without a prudential judgment similar to the one we employ when we determine if the data we have are sufficient. There is no mechanical or absolute procedure in order to know whether I am taking into account all the relevant facts. In other words: in order to determine the amount of events or facts in a percentage is something which is determined by a prudential judgment, and as Von Wright says, this does not happen without reference to one's own state of knowledge or ignorance.

Lastly, von Mises and Reichenbach believe that probability means certain frequency which is empirically observable. But this theory suffers from a misunderstanding of concepts. Probability is perhaps (and not always) *measured* by a certain frequency, probability may be the cause of a certain frequency of events, probability may be *inferred* from a certain frequency observed, but no probability has frequency as its *meaning*. First, we pointed out that physics often measure intensities of spectral lines to calculate the probability: if the line is brilliant, the transition of state is highly probable; if the line is cloudy, the transition is slightly possible; if there is no spectral line, the transition is not probable. Now, these intensities are not frequencies by any means. In addition, there are other ways in physics to measure probability, which are not frequencies. Therefore, it is absolutely false that probability means frequency.

But the most important thing to point out is that probability is probability of a *possible* event, that is to say, an event that does not yet exists, while the frequency —in order to be an empirical data, as von Mises and Reichenbach would want— is the computation of the event that have occurred, events that exist or that have existed. Perhaps the observed frequency authorizes us to *infer* a certain probability for the future, but this demonstrates that probability is *not* frequency, for the former would still have to be inferred while we already have the latter. One doesn't infer A from A; if we have A, we do not need to infer it.

By the way, the inference in question requires as a premise a principle which is itself unverifiable and which is must certain false 'there is regularity in nature' and 'the future resembles the past'. It requires it because the rationalizing assumption of the inference is that the frequency of the past will also be the frequency of the future.

The individual facts of which we speak when talking about frequency or probability are distinct: in the first case we say that from one hundred observed events, x *were* positive. In the second case we speak of a new event which hast no yet occurred and which is *not* one of the one hundred cases that have been observed. How justified is it to speak about the probability of such *future* event after proving something in regard to one hundred different events is a question we do not need to go into now. In any case, it supposes a highly doubtable premise, which is the so-called principle we previously alluded to. But even if we graciously supposed that this logical step is valid, it remains clear that frequency and probability are different concepts. First: because they cannot be predicated simultaneously in regard to a same event. Second: because in order to go logically from one to the other one needs the intervention of a highly metaphysical principle: a clear sign which tells us that the content of one of the concepts is not the same as the content of the other.

6. Life

If what has been said in this book proves to be right in its thesis that physics is not an empirical science, with much more reason can we say the same in regard to biology and behavioral disciplines, since their own cultivators only refer to the empiricity of the physical sciences as the ideal their methodologies strive for. Nevertheless, we need to explain the non-empirical character of the biological and behavioral sciences, for the belief in their empirical character is widespread and has particular consequences.

Hegel expressly warns about this: "That which is alive is an example of what cannot be understood by the abstract intellect" (PR III 71). It is important to have in mind the difference that we already established between: reason (*Vernunft*) and the abstract intellect (*Verstand*).

Proceeding like this intellect does, the biologists use words to which they cannot give any meaning and therefore they do not understand them. I will deal specifically with three very important topics: 1) Life and organisms; 2) Normality or abnormality, that is to say, health or sickness; 3) Species.

If Biology does not succeed in defining empirically that which is alive, it does not even limit its own field of investigation. Now let us look at what Baker and Allen admit in the name of all biologists: "there is no concrete line between what is alive and what is not" (1970, 3). The same acknowledgment is found in the *Britannica*: "There is not as yet a set of nonarbitrary characteristics that mark the distinction between living and nonliving systems." (EB 25, 684, 2).

We would like to insist in this point, since the common sense frequently believes that there is no problem here, and that belief is based on the false security with which the biologists think that they can proceed undisturbed, even if they do not really know what they are speaking about. It is easy to the common sense to say: all that moves is alive. But the sea moves, the wellspring moves, the volcano lava moves, the flame moves, the sun moves, the river moves; and no one, not even the common sense, affirms that these things are living beings.

Nutrition, growth and reproduction are the three characteristics most commonly considered to be distinctive of what is alive. But it is difficult to define these three terms in such a way that they do not apply to inorganic objects as well. Of course, one tries to define them by means of empirical data, and under those conditions nutrition and growth are almost the same: local transfer of external matter to the bundle of the studied object, that is to say: aggregation or incorporation of external elements of the body we consider to be alive. Nevertheless, that growth or aggregation of new elements is also observed in crystals, which are not organisms. It can also be observed in the flame, which undoubtedly nourishes itself and grows. Reproduction has two different kinds of problems. On the one hand, reproduction is not a sufficient criterion to define life: under that consideration neither the ox, nor the working bees, nor the human eunuchs nor children could be called living beings. In the case of the first three, one could not even say that they can reproduce themselves 'potentially'. And on the other hand, the flame not only nurtures itself from external elements and then grows, but also produces other similar flames and in that sense it reproduces itself. And if we wish to add as a fourth characteristic the dissimilation or elimination of the waste material, we would still face a lot of problems, for the ashes are the waste of the flame.

But the difficulties do not finish there. Each cell of the organism holds these four characteristics, and Biology would remain disabled to decide if a dog is one or many living beings.

Therefore, it is necessary to make a fundamental consideration.

If the finality seems to modern scientists something brought from the outside and not empirically evident, that is because they believe that the very object is empirically evident and they figure that by the sense data they can determine whether they have or not a living being in front of them. All the absurd pride they take in their contempt of the teleological is grounded on the supposition that the terms they use have empirical meaning and that is why they the task of knowing the object was of primal importance to them. But none of the empirical features that we have enumerated allow them to know if a dog is a living or multiple living beings, and nevertheless biologists and we *know* that the dog is only *one*. The question is: How do we know that? It is impossible to know the unity and non-multiplicity of a living being if we do not follow the consideration that the cells contribute to the wellbeing of the whole organism as an *end*. In determining its object of study, Biology uses of Teleology, despite the fact that it claims the opposite.

Moreover, the growth in the empirical sense of the mere amount of matter is an absurd criterion, since there are tumors and odd formations. For instance: a deformed head that is much bigger at one particular side —something which is not considered beneficial, but rather a very negative condition that leads to death. The incorporation of more matter only seems like life to biologists if this contributes to the wellbeing of the organism.

"The Aristotelian concept of nature is superior from that one used nowadays, for what is essential in the former is the determination of the end as the internal characteristic of the natural object itself" (GP II 173).

The difference between Aristotle and modern biologists consists in that the latter do not think enough on what they are doing; they do not notice how they are thinking.

Aside from what has been said before —which cannot be put into question—, there is a special point in which Teleology cannot be disregarded: the reproduction in the animal and vegetal realms is unaccountable if the *species* at itself does not act as an *end* and *exerts thereby causality*. Hegel highlights this: "the species is end" (GP I 381), "the universality or species is the inner" (WL II 435).

One speaks in vulgar zoologies about certain kind of insects whose male dies after the coitus: he leaves his own member and bowels inside the female's genitals. Scientific treatises supply even more dramatic examples. Let us only mention a few. Entomologist Ashley B. Gurney speaks of a particular kind of orthoptera:

> A striking sequel to mating occurs frequently in mantids when the female eats the male. There is a popular opinion that mantid males always are eaten, but many escape under natural conditions. But in the close confines of a small cage cannibalism of the male is more common. (EB 21, 607, 1).

Let us understand this correctly: when the male decides to mate, he is almost certainly heading towards his own end. Among animals the individual is not the end; the end is the species; and this happens with real effectiveness, not only on the theorist's minds.

According to the zoologist Justin W. Leonard, in most of the ephemeroptera species, both the male and the female die shortly after mating and safely securing the fertilized eggs (cf. EB 21, 602, 1). This means that, when the couple 'decides' to copulate, what they are deciding is to die. In the animal kingdom there is no individual instinct of conservation which can oppose the good and wellbeing of the species. When some atheist thinkers say that we can find some acts of altruism among animals —as in the case of the paternal care of the offspring, which jeopardizes even one's survival—, they forget that, in order to have an *alter* one needs to have also an *ego*, something that does not exist among animals. As a matter of fact, among animals the true unity is the species.

The zoologist George C. Kent tells us the following about the reproduction of the nematode (some wind of worm): when the ratio between the males and the females of a group is not optimal, then all the individuals will have a sex reversal (cf. EB 26, 684, 1). The same happens with certain species of mollusks and insects. In a word, the individual is not respected: he is in the world for the sake of the species and not *vice versa*.

In the barnacles —which are, by the way, hermaphrodites— occurs very often that normal sized individuals do not produce enough sperm; when that occurs, their male offspring are all dwarfs; all the parts of their bodies are stiffen and stingy, with the exception of the testicles (cf. EB 26, 685, 1). Whenever the evolutionist biologists describe —with the particular emphasis that characterizes them— the selective mechanism by means of which only the fittest survive, they provide us with a tautological expression. All that the Darwinist wants to say is this: the fittest for survival survive. In the light of the case of these degenerated barnacles we must say: we are not dealing with the fittest ones for survival, but rather with the fittest that can make survive the species. In the case of bees, "the male lives only briefly, just long enough to mate" (EB 21, 661, 1); males "live only for a short time at a specific time of year" (EB 21, 661, 2). And we find even a more extreme instance among unicellular organisms than that which we mentioned before: the whole male individual transforms itself into a gamete and the rest of his body disappears (cf. EB 26, 656, 2). The materialistic explanation of evolution cannot be tenable, for it is entirely based on an alleged tendency of self-conservation of the individual material bundle. After all, even stones outlast the pass of time. The Darwinian mechanism consists in that the species survives, *because* the strongest and fittest animals for the *individual* surviving, are those that have the bigger chances to procreate, while weak and feeble animals are either eliminated or excluded from the possibility of mating. The above mentioned examples demonstrate that this is not the fundamental mechanism of conservation of species. The effective leadership that the species carries through cannot be explained materially.

Let us mention only some few examples of total subordination of the animal individual to the reproduction; this time is about the female. Among certain kinds of water flies (dipteral), reproduction is parthenogenetic; the eggs are fertilized and developed inside the mother, and when they are big enough they "escape by destroying the body of their mother in a process called *paedogenesis*" (EB 21, 589, 2). In the so-called ephemeral fly, or ephemoptera, which is fertilized by the male, the delivery is sheer suicide: the female lets herself fall over the aquatic surface from a considerable height, and with the impact all the fertilized eggs are expelled out of the bowels of the mother, who dies right in the moment when that happens (Cf. EB 21, 602, 2).

Few things are more certain in Biology than the above quoted sentence of Hegel: "The species is end" (GP I, 381).

One needs only to read carefully any modern treatise on Biology to see how the very necessity of understanding objectively the phenomena makes the treatise writer forget the antiteleological dogma which he had imposed to himself in his methodological statements. A competent physic can bear witness of what I am saying, for his professional sensibility discovers immediately finalistic terms and considerations which do not occur in physics. Niels Bohr says: "Actually, we must recognize that the requirements of objective description, in tendency at least, are fulfilled by the caraceristic complementary way in which arguments based on the full resources of physical and chemical science, and concepts directly referring to the integrity of the organism transcending the scope of these sciences, are practically used in biological research." (1987, 76)

We could make an anthology with quotes of biologists who only by means of teleological considerations can seize the objects of study. Let us see some cases of this, in particular, the case of N. J. Berrill, an outstanding specialist of sexuality in the animal kingdom:

> This division of labor between mating types, male and female, respectively, is nature's way of attaining two ends. These are the bringing together of the gametes so that fusion may take place and the accumulation of reserves so that development of a new organism can be accomplished. The first calls for as many motile cells as possible; the second calls for cells as large as possible. These different requirements are practically impossible to satisfy by a single type of cell. (EB 27, 247, 1)

This paragraph does not need any further commentary.

Another instance of this source is to be found in Baker and Allen, in spite of the fact they affirm that "modern biology is based upon de presumption that the living system's functions can be explained on terms of chemical and physical processes" (1970, 2). However, when they speak of the placental mammals, they say the following:

> The *placenta* is an extraordinary organ. At the same time is an intestine, a pair of lungs, a kidney, a liver and an endocrine gland. Many substances that go from the mother to the embryo pass through the *placenta* by active transportation [...] The placenta does not work for the child's wellbeing only. It also secretes hormones which charge the maternal mammary glands to bring them to a condition in which they are capable of secreting milk (*ibid.* 429).

Each time the treatise writer says this 'works to [...]' he is doing Teleology, and confirming, involuntarily, this thesis of Hegel: "the good, which is by itself an end, is also a principle in the philosophy of nature" (GP I 472).

In regard to the lack of conscience of biologists, Hegel makes a remark that, from the point of view of the epistemologist, is of primal importance: that lack of conscience "remains attached to the determinations of the imagination, such as *impulse, instinct, necessity*, etcetera, without asking what these determinations are in themselves" (EPW 359 A).

Along the tree concepts mentioned in this text, one must add other two which are closely intertwined with them: *pain and pleasure*. Without this set of concepts, not only zoologist cannot go any further, but behaviorism proves incapable of granting a meaning to the principal term of this discipline: reward, gratification and stimulus.

In regard to the case of pain, Hegel remarks: "living things have over the unanimated ones the privilege of pain" (EPW 60 A).

Their already alluded lack of consciousness consists in believing that the above mentioned terms have empirical meaning, when only the reflection of the subject over himself can grant them meaning. We said before that not even the tendency to eat, that is to say, the impulse or instinct that we call hunger is an empirical data. All the attempts to ascribe meaning to a physical fact which we can perceive by means of the sight and the touch are doomed to failure. For instance, to believe that the meaning of hunger is the act of eating is tantamount to say something stupid, for we can eat without being hungry. The lapse of many hours in which we do not eat is not the meaning either, for those who have carried out diurnal hunger strikes bear witness that the sensation of hunger fades away after some time. And not even a single exterior sign as starvation means hunger because, on the one hand a fat and healthy person can have hunger, and on the other hand, someone could be starving because he has been beaten up or is simply sick. Besides, not even gastric secretions can be pinpointed as the physical meaning of the word hunger: like every gastroenterologist knows, one can segregate an enormous amount of gastric juices due to different worries and anxieties.

I do not believe it is possible to define impulse, instinct, necessity, pain or pleasure doing without the idea of 'end' —let alone to explain the existence of this entities. But precisely the introspective character of such terms brings us back to our point of departure: the concept of life.

As we have seen —and this something accepted by the biologists themselves —, not a single or a set of empirically perceivable data is the meaning of the word life. Now (cfr II, 2), it follows that this concept does not have an empirical origin, since its cause would have to be some empirically perceivable data. The inevitable conclusion of this is that the concept itself of life was originated in self-consciousness, and that its primal meaning is a reality known by self-consciousness. When someone said 'life' for the first time, what he said was: like myself. Life, in its fullest sense, is the self-determination and -causation of the spirit. We predicate life of plants and animals in a diminished and deficient sense. As Hegel says: "his own self-consciousness is that which man makes objective to himself" (PR II, I 94, 1); "in so far as the mind thinks of itself as changing but remaining simple on that change, it thinks *life* as such" (GP II 452). The 'salient point' in the concept of life is "the moving of oneself, the direction of oneself as such" (EPW 359 A).

In order to understand the next paragraph, one must keep in mind that the concept is the spirit, and by reflection, in its derogatory sense, Hegel means the activity of the abstract intellect with its mirror games (Cf. V, 3), since they do not add any content at all.

> The omnipresence of the simple on the multiple exteriority is for the reflection an absolute contradiction; and insofar that it is forced by the perception of the life to understand such omnipresence and to concede the reality of that idea, it results to him an *unintelligible* mystery, for the said reflection does not understand either the concept or the concept as substance of life (WL II 416). [...] It is from the *idea of life* that the idea of spirit has issued. (WL II 435).

True life is self-determination. Biologists could deny this concept only if they had another one. But we have seen that they do not have another one.

There have been many *snobbish* people who pretend to get indignant about the traditional judgment that catalogs animals and plans as inferior living forms. But in order to make meaningful such a grandiloquent statement, the person in question would have to define life in a different way than Hegel. But he or she is unable to do that.

The Aristotelian definition of life, *motus ab intrinseco et in intrinsecum*, only comes about fully in the spirit, since in the inferior living beings some parts are exterior with respect to other parts and hence movement is not entirely intrinsic. At the end of the day, however, movement does not originate inside of animals, but in the agent who put inside them such instincts and tendencies; we would not say that a toy moves itself. What moves the toy is the person who sets it into motion. On the contrary, the decision to move can go in man against the impulses and the natural inclinations; the origin and the last instance of movement is the self itself, and the determinations that are produced thereby are the being itself of the spirit, the intrinsic.

> Life of animals —although, as life, it is an idea— does not represent yet the infinitude and freedom itself which is only manifest when the concept permeates its adequate reality so fully, that only has itself, and it makes no other thing that producing itself (Ästh I 229).

Animals and plants have life insofar they resemble man; insofar they, in a gradual way, possess self-determination and move themselves. A horse moves itself —although this is not very intrinsic—, for his legs are exterior with respect of the rest of his body.

> It is true that mind has before itself a general image of the vitality and of its organization; but in real nature this general organism is dividend in a realm of particularities, from which each of them has its own limited type and figure and their own particular degree of development (Ästh I 227).

"The *inmediacy* of the idea of life consists in that the concept does not exist as such in life; therefore, its concreteness its submitted to multiple constraints and circumstances of the external nature and may appear in the more measly forms" (EPW 368 A).

The range or scale is very wide, precisely because the life one predicates of plants and animals consist in 'resembling to [...]', and that every resemblance says 'more or less'; gradation is inherent to the very concept of resemblance. Hegel would have seen a confirmation of his thesis in the fact, recently discovered, that viruses —the lowest degree of life we know from— is sometimes organic and sometimes inorganic, a phenomenon which perplexes biologists nowadays.

In addition, this gradation is the epistemological root of the judgment of health or sickness, normality or abnormality, in regard of which the naivety of the empiricism is portentous. The biologist, the physician, and even the psychiatrist believe that they are making an empirical verification when they say that certain specimen or certain behavior is abnormal, but that is in fact a judgment of value —that kind of judgments which were apparently banished by the empirical sciences.

The key is that the above mentioned professionals use judgments of value that are commonly accepted ('healthy', 'sick', 'bad functioning'), and they do not realize the nature of the judgments they draw, because everybody agrees with them and hence they infer that they must be empirical data. But how could that happen if, according to them, humans could only agree about empirical data! The judgments in question have an evident evaluative nature, but it is trivial, uncontroversial and, for the same reason, unapparent.

In their best attempt to reduce normality and abnormality to empirical data, the above mentioned disciplines say that normal is the average, that quantitative data which is available for anyone who knows how to count. For instance, the abnormal functioning would be that which entails death before the organism in question reaches the average longevity of its species. However, it is obvious that longevity does not constitute a good functioning when it comes along with never-ending pain. The existence of painful, chronic diseases is an undeniable fact. Another misguided recourse to the quantitative would be to call healthy and normal what is standard among a population, but every responsible physician would contradict that criterion, not only because it is difficult to find one single country in which the majority of the population lacks sickness, but fundamentally because the concept itself of sickness resists against that reduction to the quantitative, a reduction by means of which one would declare —with a reckless apriorism —, that it is *in itself impossible* that the majority (or even the totality) of the human kind became infected by a dangerous and contagious plague. A medical science that reacts by saying that there is no care because the majority has it, would have lost sight of the importance of his mission and even conscience.

Psychologist Abraham Maslow says this correctly: "It is average in our society to have a sick, pathological sexual life (from the psychiatric point of view). This does not make it desirable or healthy. We should learn to say average when we mean average." (1970, 266s).

'Healthy' and 'more 'frequent' are not synonymous concepts. By a simply decree we cannot exclude *a priori* the possibility that sickness is more frequent than health.

As a matter of fact, despite it deceives itself by stating the contrary, the criterion of frequency is not the one employed by the medical science in order to judge whether an organism is healthy or sick. First, this science judges that a certain group of individuals are healthy, and later seeks a sensible data that such individuals share as a common denominator, and once this is found, it uses it as practical and operative criterion of health. This is why some figure that the judgment of health or sickness is in fact an empiric judgment; but in fact the election of this criterion depended on a previous judgment that was not empirical. If physiologist would have found in the first place that individuals with seventy-two palpitations per minute suffered from chest pain and breathed with difficulty, they would have not adopted this quantitative data as criterion of 'good functioning'. The key question is the following one: how would they know, by empirical data, that the average number of palpitations *constitutes* a good functioning? The fact that the majority possess that number of palpitations is not a synonym of health, since the majority could be sick. First, medicine assumed that the majority was healthy, and it did that gratuitously, that is to say, by means of a judgment of value whose nature remained unveiled because the rest of people shared the same thought. As we said before, those are trivial and uncontroverted judgments. Only *because* the majority believes to be healthy, medicine takes the criterion of what the majority believes. In fact, however, the medical science does not exclude the possibility that a population that considers itself to be healthy is actually sick.

Despite, the impetuosity of those biologists who claim that health is an empirical data, we have seen that that is not the case. One could ask then, what is the origin of the judgment of health?

'Vital' in its original meaning, is the thing that entirely determines itself —which is the spirit. It follows from this that 'less vital' (*i.e.,* sick) is what is being determined, in higher or lesser degree, by outer agents which reduce its self-determination, liberty or capacity of moving to itself. The lesser point of reference is the inorganic, it is to say, the 'object' that is externally determined by others and has no initiative whatsoever. The organism is reduced to that state in a higher or lesser degree by sickness.

The same gradation that allows us to affirm life in plans and animals —insofar they resemble more or less the self-determination of the spirit— constitutes the possibility of judgments of health and sickness. The self-determination of animals and plants consists in that the specificity of their reactions, forms and characteristics is not determined by the environment but by the hereditary germ which each species possess. This makes us see that it is not a full self-determination, for it is not the individual organism the one that decides but rather the species, that "universal individual" as Hegel called it (PG 218). However, even the limited and flawed self-determination decreases in case of illness, and illness should be defined by the diminution whose extreme possibility is suppression (death).

Let us note finally that the same gradation of the concept of life —whose concretion, Hegel told us, "is subjected to multiple conditionings and circumstances of the external nature" (EPW 368 A). The gradation rules the formation of the concept of the diverse species of the organic.

The multiple attempts that Biology has made in order to determine the species on the grounds of physical and empirical features have failed. That was the case of a person named Blumenbach, who in times of Hegel affirmed that the earlobe was the essential feature of the human species. Biology could have spared too many troubles had it read carefully what Hegel said about such physical features:

> By their exteriority one notices that the knowledge of concept did not originate in them: it was an obscure premonition, an indeterminate yet profound sense, a forewarning of the essential, what preceded the discovery of the species in nature and in the spirit, and one only searched afterwards, for the abstract intellect, some determined exteriority (WL II 456).

It has been a delusion, an illusion of the empiricism like the one we mentioned apropos the judgments of health and sickness.

For responsible biologists it is nowadays evident that their discipline cannot justify by empirical data the formation of their concepts of species. As Munson and York report, "some biologists have concluded that species have only a subjective existence merely as convenient labels for arbitrary assemblages and have only a minimum of biological significance" (EB 25, 686, 2). Furthermore, Herbert H. Ross, one of the most acclaimed specialists in taxonomy, makes all the more evident this spectacular failure: "We might find that different populations each previously considered to be separate species are only one, or that different populations previously considered to be a single species actually represent many species." (1974, 13).

For centuries, in an effort to set their concepts of species in empirical data, Biology drew its attention to describable features and characteristics. But this attempt failed innumerable times in facts like the following one: some butterflies have only four legs in contrast with the model of six legs that characterizes almost all insects: having six legs could not be longer considered any longer to be a particular empirical feature of insects, because on the 'basic' and 'fundamental' level, butterflies had to be considered as insects. This was the crack that made the entire building of the anatomical criterion fall down: to call some features essential and other not is a judgment of value and not an empirical data. To justify such a judgment by empirical observation is impossible, for this judgment determines precisely which observations are basic and fundamental and which observations are not. We could quote many examples like this.

In the light of this failure, biology —not willing to renounce to its status of empirical science— employed the criterion of mating and intraspecific fertility: it decreed that species is characterized by the fact that their individuals mate between themselves and have offspring. But this failure is just as spectacular as the past one. First, many species reproduce themselves by simple fission, others by parthogenesis, other species are hermaphrodite, and so it turns out that we cannot divide animals in species by look only to the fact of the mating of the individuals; the above mentioned types of reproduction —more particularly, the first one— are quantitatively very important in the animal kingdom, and in those realms the term 'mating' does not have any meaning at all. Second, if one adopts this criterion, one must abandon the previous one, namely, the description of the essential characteristics. Biology, however, does not come to terms with this idea. Baker and Allen tell us: "Two progenitor plants, which are capable of producing a hybrid offspring, are eventually considered as distinct species because they differ in their anatomy and in other important details" (1970, 462). Biology works ludicrous conceptual tricks to feed its own complacent delusion of being an empirical science.

Third, the objection of Benjamin Burma is tremendously powerful:

> What, then, is a species? It would seem thus far to be the whole of any one series of breeding populations... [But the] definition as it stands unfortunately puts all living and fossil animals in one species, since there is a continuity of germ-plasm back from John [an individual animal] to the original primordial cell, and from it forward to every living animal (not to mention plant). (EB 25, 686, 2)

In order to reply to Burma's objection, one could define species as the set of populations that have intercourse at a period of time, but then the number of species would be infinite. There would not be a temporal continuity among a species, which is something utterly absurd.

The origin of all this mess was to believe that the concept of species is obtained a *posteriori*, i.e. by generalization of all the particular cases. But from a logical or an epistemological point of view this is impossible. How could we know from which individuals —among the many existing ones in the world— should we abstract the concept in question, if we do not observe the world with the concept that will determine our selection? By means of which criterion can we rule out the sets of animals that are not useful for us? By means of which criterion, if is not the *a priori* concept, can we rule out the monstrosities and hybrids, that the very experience displays us?

> The trouble lays in the impotence of nature to hold the concept in its verification [...]. Nature revolves everywhere the essential boundaries, which always present new instances against every firm decision, even within a determined species (e.g. men), by means of monstrosities that, on the one hand, must be ascribed to the species in question, and on the other hand, lack the determinations that must be regarded as the characteristics of the species. In order to consider such forms as defective and deformed, one must suppose a fixed type, but that cannot be collected from experience, for experience provides us also with such monstrosities, engenders, etcetera; a fixed type supposes rather the autonomy and dignity of a concept (EPW. 250 A)

"It follows from this that only *life in general* can be valid for the observation of the concrete forms, but when that life is fragmented it does not have any order or rational classification in itself, it is not a roundabout system of forms" (PG 219). The concept of *life*, as we have said, is obtained by means of self-consciousness; it is *a priori*, not *a posteriori*. Its content is the self-determination of reason and of the spirit: that is *life in general*. From plants and animals we predicate life in a deficient and diminished sense, because such a realization of life "is subjected to many conditionings and circumstances of exterior nature" and non organic. Each species is life *insofar* it possesses such material conditionings and particularities; we obtain the concept of species by means of selection, that is to say, by selecting the true content of life according to the possibilities that the material world offers. One species crystallizes in front of our eyes —so to speak— only when a picture of unconnected points suddenly appears to us, as in the gestalt experiments, as a figure that 'makes sense'. In our case, to have meaning means that it is viable and can have life. One should not be surprised that such configurations called species have a provisory character, as the taxonomist Ross warned us. We maintain that species only insofar that there is not instance that contradicts it. We maintain it because it has sense, because it is a possible form or realization of the *a priori* concept of life.

7. Conduct

Although Hegel could not foresee the contemporary boom of the concept of behavior, such notion falls undoubtedly under the judgment that he left Hegel formulated: "Life as an example of what cannot be understood with the abstract intellect" (PR III 71). After all, a behavior is a piece of life, one piece among the many that conform life, one unit of that which we call vitality. If only by means of self-consciousness is possible to grant meaning to the term life, the same must happen with the term behavior.

"In the empirical reality each action has many precedents, so that it is very difficult to determine in which point is the beginning" (Ästh III 274).

In the search of *a* behavior we find the same gestalt procedure we just mentioned in regard to the concept of species: a behavior is a vital unity that 'has sense'. We could not justify by empirical data the conceptual selections we make in the temporal *continuum* in the life of an organism in order to affirm that there is *a* behavior between the two cuts. The rat does not cease to move once it eats. And, in fact, the rat was not still before. No sensible data tells us: here a behavior ends and here another begins. To be sure, there are empirical data: but none of them *means* 'limit'. The criterion by means of which we will judge the ulterior movement does not belong to the behavior to which we paid attention before: evidently, it is a question that is not determined by empirical data. The temporal *continuum* which is the life of the observed organism is not empirically sectioned in parts. Our verbs of action —projected over the said organism— constitute different types of unities. The subject experimented as a unity certain segment of his own existence and named that set of movements. Only afterwards he projected those unities and the cuts that limit them over the other organisms. It is only due to the interiority of the subject that such sets have unity.

I am not saying that we do not know when a behavior ends and another begins. Of course we know that, but not by means of empirical data. If the observer remains in the empirical, he could only speak at best of movements, but not of behaviors. Behavior says something much more that movement: it means *activity*. Now, if some concept was originated in self-consciousness, was the concept of *activity*. It implies causation, and we have seen (III 8) that causality is not an empirical data. One says that he *behaviors* himself in one way or another, only because he realizes that it is him who determines the course of his actions. If that was not the case, he would not behave himself he would be manipulated by something else. Only in a deficient and derived sense we can say that animals behave themselves: "Both the action and the event are originated in the interiority of the spirit" (Ästh III 139), "the agency begins with subjectivity" (PR II 211).

It is amusing that behaviorism —grounded itself entirely in an introspective and 'mental' concept— criticizes all other psychological methods, accusing them of mentalism and anti-empirical.

If the behaviorists decree that 'behavior' is movement 'for them', they are going directly to a death end. In fact, a behavior comprises many movements, but the important questions are: How many? With which criterion does the observing mind comprise them as a unity? On what grounds does the observer affirm that many movements constitute only *one* behavior and not many?

For instance, how could we know in a banquet if I had one or several behaviors? If I drank one glass of wine in five sips, did I have one behavior or five? If the rat ceases to eat while eating, would we say that it carried out as many behaviors as the times it stopped moving?

Two extreme behaviorists, Tolman and Guthrie, addressed this subject, but they displayed an astounding lack of reflection while dealing with the epistemological problem that it entails. Winfred Hill describes this very accurately:

> Walking a city block, for example, is a molar act made up of an enormous number of molecular movements —expansions and contractions of the various muscles of the legs and other parts of the body. Guthrie is an example of a theorist who puts a good deal of emphasis on molecular analysis. Tolman, on the other hand, states explicitly that he is concerned only with molar behavior. The ways in which molecular movements work together to produce molar acts are of no concern to his system. (1983, 133)

We must say something in regard of the tastes and preferences on account of these investigators:

First, Guthrie himself falls short in his analytic preference. For instance, biologist Frank Brown speaks of "This is well illustrated by the complex movements of swallowing in mammals; in the dog, for example, 11 separate muscles or muscular systems are found to discharge one after the other, precisely timed to a matter of milliseconds". (EB 14, 636, 2). According to the preferences of Tolman, one behavior would comprise a long series of events. On the contrary, Guthrie would say that each single movement is a behavior. Furthermore: if one tries to reduce behavior —doing without its genuine meaning— to physical movements, why should one exclude the atomic movements of each muscle from our analysis?

Second, if it depends on the taste of the observer the number of behaviors we are dealing with, it follows that the object of study of behaviorism is not an empirical data. The criterion by means of which the investigator builds up a behavior is not extracted by empirical data but from his own understanding and self-consciousness. Therefore, behaviorism is no other thing that a reloaded mentalism.

Against the above mentioned decree, we would like to pose a dilemma which seems to us to be definitive. And this, of course, has to do with still behaviors. But before we formulate it, let us see some other problems first.

Will behaviorism deny that a paralytic has a behavior? Should we say that a disabled person that pays us attention does not have any behavior at all? Is hearing not a behavior? Is looking carefully instead of being distracted not a behavior? If I stay still during five minutes while listening a *sonata* with attention, could someone say that I am not behaving myself somehow? Is listening to a *sonata* not a behavior?

These still behaviors can have tangible effects. For instance, a forester can contemplate attentively a fire during five minutes in certain direction, and then make a phone call in which he says that there is no fire at the place he has to watch. If it has some undeniable empirical effects, would the behaviorists deny that such still behavior existed?

And let us go now to our dilemma. Against the behaviors we have mentioned previously, behaviorism can assume two positions:

First, behaviorism can acknowledge that there are in fact behaviors. In order to do so, it would appeal to the (supposed) movements of the neurons inside the head of the agent. But then the circulation of blood would be a conduct which, by the way, is not motivated by an exterior stimulus. That would be a decisive proof against the behaviorist theory, for it pretends to explain any conduct by exterior stimuli, and ignore the mentalist entities. The recourse it could employ is to distinguish the behaviors of the paralytic and the forester in one hand and in the other the circulation of the blood pointing out that the firsts are in terms of voluntarism, and the second don't. However, if that happens, the will would be an essential element of conduct, and few things are as introspective and mental as that!

Second, behaviorism could stubbornly affirm that the allegedly still behaviors are not behaviors because they are not exterior movements of an organism. If that were the case, the salivary and gastric secretion of the Pavlov dogs would not be a behavior. We should not forget that the entire behaviorism originated in the observation of that very fact as some kind of paradigmatic behaviorism. To be sure, behaviorism could try to escape from that *reductio ad absurdum* by decreeing that we are dealing with a behavior when we consider the entire organism as a whole, but the salivary secretion is evidently not a movement of the organism as a whole. Furthermore, there would not be any conduct at all, since not even eating is a movement of the organism as a whole; my Saint Bernard dog i.e. does not rise from his place to eat when I bring near him his plate; during his meal, most of his body continues exactly as it was before.

Each and every one of the above mentioned alternatives means the fall of behaviorism, unless it relinquishes the thesis of behavior as a physical movement. But then behavior is an introspective concept, and that would also mean the fall of this theory.

We have not talked about the word stimulus. It is obvious that this word is meaningless if does not refer to different impulses or instincts, or needs, or to pleasures or pains, being all of them metaphysic entities.

In order to avoid the proper meaning of the term stimulus, behaviorism would have to arbitrarily decree that 'stimulus is every empirical fact that *explains* the existence of a behavior', for the epistemological status of behaviorism is eminently explanatory. Its primary intention is to formulate laws (cf. v. 4). Let us not bring into consideration what we said before in regard of unempirical terms like 'always' or 'every', without which no law can be formulated. The sole reflection upon the matter would show how frustrated their attempts are. What is specifically important to point out is that without the consideration of an interior impulse or a pleasure, the definition of stimulus above mentioned would not *explain* the existence of a given conduct nor *make it exists*. We are not witnessing a behavior when we apply the same stimulus to a brick. Therefore, the stimulus taken alone does not make that behavior exist. Something more is required in the body we are observing: some instinct, necessity or pain: all those things are mental. In spite of the arbitrary recourses it employs, this arbitrary definition cannot avoid the internal.

To make matters worse, behaviorists themselves cannot deny that behavior can occur without the stimulus whose effectiveness they study. It is very well known, for instance, that the movements of the mouth of an infant or a chimpanzee who want milk while asleep are a behavior without the belonging stimulus. Then what is missing is to reflect on the epistemological status of behaviorism, and this would make its existence to be justified. For example, (and we could talk of many others): It is not the stimulus that brings behavior into existence. But, if behaviorism is not explanatory, what would be then its status? It is not descriptive, because terms like reasoning and stimulus, just to mention a few, are evidently explanatory.

CHAPTER VI

The Man and the State

All fundamental mistakes in political philosophy and in the social and human sciences (including the legal theory of Kelsen) stem either from one or two of the following assumptions: the first one is to believe that man is naturally good or even that he is man by nature; the second one is to consider the studied entities as empirical data. I am referring to entities such as society, state, freedom, lack of freedom, language, equality among man, the Right, The Law, humanity itself, customs, behavior, the authority or government, property, production mode, etcetera.

Since we have previously demonstrated several important theses about man, we strongly advise against reading these chapters without reading the previous ones. Following our analysis, it is of primal importance to consider that not even the entities of Physics and Biology can be discovered by appealing to empirical perception. After this meticulous study, it would be childish *to affirm* the equality among men —in the belief that it is an empirical data— without previously demonstrating it. The empirical data is the inequality of men: some are bigger and others are smaller, some are fat and others are thin, some are active and others are passive, etcetera. Nevertheless, if equality is not demonstrated, our unavoidable demand of democracy is not justified either, and anyone who defends it would have as much reasonableness —or unreasonableness— as those who defend autocracy. This subject will become a question of 'taste' outside the margins of rationality. The dispute among men about political theories will cease when the truth of them become an obligation to be demonstrated.

And if we are talking about demonstrations in politics, surely Hegel's political theses have a lot to say to us. What we have been saying does not pertain only to the problems themselves but also to the interpretation of Hegel. He takes off from the fact that he is making science and not simply boasting esthetical caprices. He states that explicitly in his prologue to his *Philosophy of Right*. If we are dealing with science, we must first define and then demonstrate. It is very funny, for instance, that people who accuse Hegel of worshipping the State are those who do not even define that concept. Those persons inevitably confuse the State with the government, despite that Hegel eagerly demonstrates they are different things. Those persons believe to be the champions of the civil society against the State, but they are incapable of defining both. They are incapable, therefore, of supposing that perhaps both things are the same, and that does not depend on the arbitrary definitions that each author has the whim of writing. If the civic society is not a physical object proved by empirical means, then it is constituted by rights and obligations. Only in that, and just in that, does the State consist.

In this present subject would be very artificial to deal with interpretative questions separately, dealing with specific matters in independent sections as we have done so far. Our exposition will be from this point onwards more discursive, except in cases like Hegel's iusnaturalism, which has been put into question by some.

One unequivocally notices that Hegel's message about man and the State, insofar it is a demonstrated truth, acquires the importance of a revelation. This is the reason why we must deal with theologians in our study too, for we are speaking about the end of man and the meaning of the entire human history.

1. Natural goodness?

Any thinking person can see that men of letters and poets attributed natural goodness to farmers and countrymen in order to idealize pastoral life. Without this element, they did not consider this landscape 'beautiful' enough to be treated by the arts. Obviously, this assumption of natural goodness is an apriorism. Unfortunately, the political left nowadays moves entirely within this apriorism. In order to be supportive with the oppressed, the indigenes and the proletariat, the left has blindly assumed that each one of its members is truthful, honest and moral. The leftist reacts with indignation when someone demonstrates him that all the felonies and crimes that the poor commit against each other are deeds as perverted as those which the working class suffers under the hands of the wealthy class. Although the rational and moral justification of the struggle in favor of the exploded ones does not need at all such romantic assumptions —which, in fact, only hinder us from constructing a world that is truly just and good—, it is very understandable that the left has felt the psychological need of employing them. The left has given much more importance to this objection than what it actually has: why should one commit oneself with a class that is not morally better than the other class? As an objection against the just revolutionary class this is useless; however, as a scientific question that demands and answer, it has a tremendous power and depth from which the left has in vain tried to escape.

Neither political struggle nor science can be grounded on vulgar, romantic reveries and argumentations. A clear example of this is rude kind of reasoning is the opinion according to which the Christian thesis of the natural evilness of man is 'outdated', only because it was formulated many centuries ago. Two plus two equals four is a truth that was formulated many more centuries ago and no one questions it.

> The Christian teaching according to which man is evil by nature is superior to that which affirms man is good by nature; it is necessary to understand its philosophical meaning. As a spirit man is a free being whose destiny is not to be determined by his natural impulses. Therefore, in conditions of immediacy and ignorance, man is in a situation in which he should not to be and from which he must free himself. This is the meaning of the doctrine of the original sin, without which Christianity would not be the religion of freedom (Rph 18 Z).

I believe that no honest intelligence can consider as a 'matter of theologians' something which in our century made Einstein say: "it is easier to denature plutonium than it is to denature the evil spirit of men" (Schilpp II 1970, 655). These words of Einstein (Conacyt revue, february 1983, p. 32) are extremely eloquent as well: "The true measure of a man is the degree to which he has managed to subjugate his ego".

Sigmund Freud was not a theologian but a very convinced atheist. Despite of that, however, he went against his deepest philosophical convictions when he discovered that every child around the age of four is a "polymorphic universal criminal". The romanticism we were speaking about has to include the child, for one understands by natural man someone who has not yet been modified by education, society or culture. Now, this apriorism collides painfully against the fact —which can be witnessed by anybody who does not idealize things— that children are consummate egoists who are capable of all imaginable cruelties.

> Against the opinions and wishes of a sick philanthropy who wanted to draw men back to original innocence is opposed the reality itself, and even the very nature of this matter, namely, that man is not destined to such nature. And in regard of children, it is in this period where their egoism and cruelty are displayed (PR II, I 33n).

In this context, there are few testimonies as evident as Rousseaunian one we mentioned before. He is the champion of the natural goodness of man: "Let us lay it down as an incontestable maxim that the first movements of nature are always right. There is no original perversity in the human heart" (Emile, II). Rousseau is the author that spread widely this prejudice. In an apparent innocuous note of that very book, however, we find the following thesis:

> So there is only one of the child's desires which should never be complied with, the desire to make himself obeyed. From which it follows that in whatever they ask for it is the motive behind their asking that must be paid attention to. As far as possible give them everything they ask for provided it can really give them a real pleasure; always refuse what they ask for out of fantasy or simply to wield authority (Emile, II)

If Rousseau had studied history carefully, or if he had lived in the times of Nazism, he would have known that the desire of domination has caused much more cruelties and atrocities than any other; and if that impulse is something natural in man, as Rousseau himself recognizes in that note, then his thesis about the natural goodness of man falls immediately apart.

In another apparently inoffensive note, we read this as well:

> I have seen silly women inciting children to rebellion, encouraging them to hit people, allowing themselves to be beaten, and laughing at the harmless blows, never thinking that those blows were in intention the blows of a murderer, and that the child who desires to hit people now will desire to kill when he is grown up (Emile, II).

Whenever it comes about dealing with reality —instead of recklessly blurting apriorisms to the world— Rousseau hits always the mark. If it were within his power, the child would kill. And Hobbes is absolutely right when he says that the savage man is simply a *puer robustus*.

In regard to the savage man, Rousseau himself refutes his principal thesis, the one that gave him worldwide appraisal, when he says that "All savages are cruel". However, the vegetarian explanation he provides for this is laughable at its very best: "it is not their customs that tend in this direction; their cruelty is the result of their food" (Emile, II). This explanation is entirely futile. What matters is the Rousseaunian testimony of a very concise fact: cruelty is something natural to man. The time has come in which we must say —regardless of those who accuse us of 'misanthropy'— that considering the savage as guilty is the only way of regarding him as a person. The sick and 'understanding' anthropology that makes pleas for him treats him as an animal, an irresponsible baby or a mentally disabled individual. Moreover, those who pamper man and make him believe that he should live 'carelessly' actually despise him. The best there is in man —an impulse towards authentic solidarity— is not given to him by nature, and one does need to live carelessly in order to follow it. It only demands that one overcomes egoism and a selfish calculus of loss and gain —things which are indeed natural.

I insist on this point: quoting Rousseau is a confrontation test and a cross-examination. Such means proves to be effective because it refutes his principal thesis.

The proverbial optimism of the Greek culture in regard to man —a culture one would never say was biased by Christianity— can provide us with another indisputable testimony. Hegel says of that culture the following:

> it originally contains joy and harmony. The Christian religion is not as joyful as this one; rather, it provokes the lack of joy; it begins with pain; it

> wakens one up and tears apart the natural unity of the spirit, it destroys natural peace. Here we have an original sin: man is evil in his origin; therefore, in the more internal realm, he is something negative in regard to himself (PR I 23).

I quoted Einstein and Freud before because they were very reluctant with Christianity. Now I will quote Plato, a true exponent of the optimistic Greek culture, in which the Christian influx is automatically ruled out. One should not believe that natural evilness is a Christian invention. It is the ascertainment of a *fact* whose importance cannot be neglected anymore by true science. We read the following in the book X of the *Republic*: "This description has carried us too far, but the point we have to notice is this, that in fact there exists in every one of us, even in some reputed must respectable, a terrible, fierce and lawless brood of desires, which it seems are revealed in our sleep" (*Republic*, IX 572 B)

Aristotle, another key figure of Greek culture, who possesses the same intellectual honesty as Plato, expressively warns against "the evil that exists in each and every one of men" (Politics 1319 a 1).

The most ancient testimonial of this fact dates back to the tenth century before Christ. It is contained in the first book of the Bible: Genesis 2, 4b; 3, 24. Hegel devoted many pages to analyze this passage and we can only do the same. First, because its depth is bigger than that of all the other testimonies we have seen before. Second, because, even though this narration originated the Christian thesis with which we are dealing, it has been and continues to be completely misinterpreted by systematic theology, as we already said in our first chapter. In this context, it is important to say that the specialization in theology does not correspond at all to the specialization in Biblical sciences, nor is it a good preparation for their study because it commonly predisposes one to read certain theses of systematic theology in Biblical passages that state other kind of things.

The first thing that our public —Christian and non-Christians— should know about the passage we referred to above is that it does not speak of an individual man called Adam. All the modern exegetists and translators know this. I cannot explain myself why the Churches refrain from telling this explicitly to the wider audience. As Hegel marvelously interpreted it— before the exegetical science had done it—, the passage does not speak of an individual man, but of *all* man.

Hegel says: "the expression 'first man' is tantamount to say 'man as such', not someone individual, fortuitous, or someone among many" (PR III 127).

"Therefore, we could say: it is the eternal history of the freedom of man the fact that he awakes from the slumber he is in during his first years and gets to the light of awareness, more precisely and in absolute terms, the fact that good and evil start to exist for him" (PR II, I 32).

The Hebrew expression *hâ'âdâm* with the article means: 'the man'. On the contrary, the expression *âdâm* without the article means a proper name like John or Robert. A similar case is found in our language in names like pearl or Pearl, rose and Rose, etcetera. That difference in Hebrew —which is expressed by the article— is the one we express by the capital letters. Now, in our passage we never find the expression *hâ âdâm*. It follows that the author wants to say something about man, about every man, about the initial processes that we all go through as human beings. This is of course some powerful evidence, something that really makes a great noise. If we analyze this with a rigorous method, we will see that it is not a sin committed by some other person, something hereditary or something that God attributes to everybody for unknown reasons. The Biblical story does not speak of a personal individual but of all men. J. Blinzer, a Catholic who is not a scientific exegetist, shows this: "We do not find anywhere some determined statement which says that the first of the first father was transmitted to their descendents" (LTK III 965).

The worst thing of all is not the mistaken interpretation that systematic theology has made, but the fact that this distraction has hindered people from understanding what the Biblical author has to say about all men. He makes two affirmations of such depth that not even the exegetists have acknowledged it; they say they are as busy as they were before in dissipating the mistakes of dogmatic theology.

First, the author, who is called the Yahwist by the exegetists, says that man makes himself, that he does not come to be a man but only by means of his own deeds and activities.

One should notice that, in order to know this, the Yahwist did not need any special revelation whatsoever: he only needed to think — just like Aristotle did six centuries afterwards— upon the fact that "the mind is nothing before it thinks" (De anima 429b 32), "it does not have actual existence before thinking" (ibid. 429a 24).

Where does the Yahwist say that man becomes by his own work to be man? In Genesis 3, 5 and 3, 22.

Furthermore, the priestly narrative of the same fact (1, 1-2, 4a) is much more abstract. It tells us that God make man in his image but he does not tell us how. That led the readers to think that God made man with the same deficient characteristics he nowadays has. The last complier or priestly writer wanted by all means to avoid that kind of interpretation; for that purpose, he simply juxtaposed the narrative of the Genesis, 2, 4b- 3, 24, which was written precisely with the intention of correcting and refuting the widespread idea according to which man was made by God as he is today; savage, egoist, etcetera.

It is not the case that the Yahwist author knew or had before him the priestly narrative that has come down to us. Although it was written at a later time, what cannot be denied is that Genesis 1, 1-2, 4a contains that creational conception which was tremendously wide spread, even if it was written in a latter period. Modern investigators are positive in saying that these ancient beliefs were not only part of an oral tradition but of a written tradition as well. They hold that the priestly writer remade some ancient document in which we have variants like 'God made' instead of 'God said', the latter being now a monotonous repetition that characterizes many rituals. Much more likely, the ancient writings did not set in a time sequence the creational activity of God within seven days. That was rather a major concern for the priestly writer, who had to justify the sabbatical rest by saying that God himself rested in the seventh day.

We can appreciate both in the final compiler and the Yahwist the intention of changing that original conception of the Creation. Many particular features reveal that purpose. For instance, according to the priestly narrative, God made the animals before man (Genesis 1, 24-25). The Yahwist narrative, on the contrary, with a highly interesting etiological purpose, says that the creation of the animals came after the creation of man. Likewise, according to the priestly narrative, God created woman and man simultaneously (Génesis 1,27); while the Yahwist narrative says " the man said: this one at last is bone of my bones and flesh of my flesh" (Genesis 2, 23), taking the creation of the animals as an intermediate stage. God did not obtain the material from the ground but from the body of man himself.

If the intention of the Yahwist —as all modern exegetists recognize— is to amend the archaic conception which came down to us through the priestly narrative, it becomes obvious that the 'resemblance with God', which makes man to be what he is, does not exist from the very beginning as the priestly narrative suggests (Genesis 1, 26-27), but only after man knows good and evil: "Then Yahweh God said, 'Now that the man has become like one of us in knowing good from the evil" (Genesis 3, 22).

Before he ate from "the tree of knowledge of good and evil" (Genesis 2, 17), that means to say, when he came out from the very hands of God, he was not yet a man. He was simply a 'living being', an expression that the whole Bible employs to denote animals. The Yahwist says with great accuracy about this very first moment: "Yahweh God shaped man from the soil of the ground and blew the breath of life into his nostrils, and man became a living being" (Genesis 2, 7) —a passage which is in clear contrast with "now that the man has become like one of us" (Genesis 3, 22).

This expression 'living soul' is employed in the entire Bible in relation to animals, just as it can be seen in the same priestly narrative a few passages before (Genesis 1, 20). The examples of this are innumerable. The Ecclesiastes explicitly says that, in having a living soul, man is no different than the animals: "they have all one breath" (Ecclesiastes 3:19).

According to the Yahwist, in order to become truly a man —in other words, in order to resemble God— man must exercise his own activity and decision in regard to good and evil. In fact, we saw (III, 7) that conscience is the cause of self-consciousness, and that being a man is to have self-consciousness (III, 1). To interiorize the moral imperative has the effect upon man of giving rise to a self which is made accountable for his actions by the preexisting intersubjectivity around him that makes him responsible. The response to that self —whether it is positive or negative— is the act by which man makes himself. As we have said before, the Yahwist only needed to reflect by himself upon the difference between humans and animals in order to discover that. We cannot project upon the first moment a kind of being different to that which we confirmed now as its characteristic way of existing and being. Hegel comments thus the narrative of the Genesis:

"What is opposed to believe that what is conceptually the man could be previously present as actual existence? That is opposed to the very nature of the spirit. The spirit is only that what he makes himself being" (PR II, I 28n). "The fundamental teaching is this: man is not a natural entity as such, he is not an animal, but a spirit" (PR II, I 27).

It is astounding that Hegel suspected directly the bottom of the Yahwist's thought, while modern exegetists can barely grasp it. The ones who approached themselves more to the thought of the Yahwist were Stoebe and Gunkel. Stoebe remarks: "In fact, man becomes like God by distinguishing between good and evil and imbedding his life with autonomy" (1953, 397s). Gunkel, on his part, said accurately: "The awakening of reason occurred with a sin against Yahweh" (1966, 31) If both commentators had remembered something that they already knew and affirmed in other contexts, namely, that this story speaks of 'man', of every man and not of any particular individual, they would have come to understand that the strong point of the Yahwist teaching is that every man becomes man by means of an act of his own. To be sure, this happens under the moral exigency that Yahweh addresses to him. According to the Yahwist, the thesis of the ancient text is still true, in the sense that it was God who made man in his image, since all the self-determination of man depended on the fact that God addressed an imperative to him; but resembling God is something that man does by means of his own activity and not something that occurs passively.

To be sure, one would be misinterpreting the genre of the texts if one raised the objection that man could speak before he ate from the tree and hence was already a man. The snake, which was also an animal, spoke as well and was not a man: Genesis 3, 1-5. Even with certain sarcasm against the lucubrations of the theologians, it is nowhere stated that the snake was the demon. On the contrary, the Yahwist warns explicitly that "the snake was more subtle than any beast of the field which the Lord God had made" (Genesis 3, 1). And certainly it is not a punishment to the devil to eat dust and crawl upon the ground (3, 14). His punishment is certainly not of that kind.

The literary genre that the Yawhist decided to employ here is the myth, which was the most widespread literary genre in other cultures to explain the origin of the world (v.g. The Epic of Gilgamesh). Now, if one picks up that genre, as Hegel says: "the inconsistencies are unavoidable" (PR II, I 30). Another inconsistency: the marriage of Cain (Genesis 4, 17) with a woman nobody knows anything about. Furthermore, what is said in Genesis 2, 17 is barely understandable to man ("for in the day that thou eatest thereof thou shalt surely die"), namely, the meaning of the verb 'to die', for he had never seen anybody die. The examples are endless.

Hegel observes this correctly:

> If the speculative, the true, is exposed in sensible figures under the guise of events, there will always be certain traces of inconsistency. That is what happens to Plato when he speaks of the ideas by means of images: inadequate situations may always show up when that occurs. (PR II, II 85).

Let us summarize what we have said: the Yahwist speaks of all men. He wants to elucidate the process by means of which every man becomes a man, and the key to that process is an act in which one decides between good and evil.

The next statement is as strong as the previous one: this first act, by means of which every man comes to the use of reason, is always a sin.

Few things have been said that can be compared to the honesty of analysis of such thesis formulated more than thirty centuries ago. Traditional theology has not had the guts to accept that teaching of the Bible: every human being begins with an act of egoism. It preferred to reduce original sin to a curious anecdote that occurred long time before, and made up a series of juridical terms in order to say that the act of that first person can be attributed to all people.

I do not know why one could doubt that the thesis of the narrative we are now dealing with is the one we have appointed, since the Yahwist himself says that "the desires of man's heart are evil from his youth" (Genesis 8, 21).

Another relevant passage is this one: "And God saw that the wickedness of man was great in the earth, and that every imagination of the thoughts of his heart was only evil continually" (Genesis 6, 5). In order to understand the writings of the Yahwist, that thesis if of primal importance, because it makes one feel that Yahweh must necessarily intervene in human history in order to change things, as can be seen in Genesis 12, 3; 18, 19.

Everybody sins when he or she comes to the use of reason; it is not the case that everybody inherited some strange sin that is alien to us. The thesis was newly formulated by the author of the book of Kings: "there is no man that sinneth not" (1 Kings 8, 46). And the Salmist says something that echoes that passage: "And enter not into judgment with thy servant: for in thy sight shall no man living be justified" (Psalms, 143, 2). We read the same in Proverbs: "Who can say, I have made my heart clean, I am pure from my sin?" (Proverbs 20, 9) In the book of Job, one can sufficiently corroborate the thesis that every man and woman commits sin: 4, 17; 14, 4; 15, 14.

Whenever theologians wield innumerable Pauline texts in order to say that sin is hereditary —in clear opposition to the teaching that sin is an act committed by each individual man— the first thing one must answer them is that they are defending the thought of Paul, not that of the Yahwist. Not only is it very dubious that Paul understood the Yahwist correctly, but it is evident that Paul, influenced by certain Jewish decadence, read a proper name where the Yahwist clearly says 'man'. As the eminent Biblical authority Claus Westermann suggests, Paul depends on the apocryphal fourth book of Ezra that says "Adam, what have you done! When you sinneth, your fall did not only befall upon you but upon us too, your descendants" (4 Ezra 7 118). Westermann clearly remarks that the doctrine of Paul cannot be grounded on Genesis 2-3. It is important to notice that the name of Adam is not even pronounced once by Christ; the conception of Paul cannot be based on the authority of Christ.

But this *second* point is even stronger: in the same passage (Rom. 5) that theologians refer to, Paul explicitly says: "for that all have sinned" (Romans 5, 12d). Neither Pauline exegetists —either leftists or rightists— nor experts in the Greek language would tolerate nowadays that the expression *ef'ho* is translated "in which". It is an explanatory or causal conjunction which means 'because'. We need only to refer to the Catholics Zerwick and Juss, and to the Protestants Zerwick and Kuss. In a like manner, they all energetically reject that one interprets the aorist *hémarton* as 'sinful state', because in reality it means act: 'to have sinned'. It is equivocal that, according to Paul, every man commits an actual sin, something which was carefully exposed in the precedent chapters of the same letter, since when Paul says "for we have before proved both Jews and Gentiles, that they are all under sin" (Romans 3, 9), the only thing he has demonstrated —by means of description— are multiple concrete sins of envies, injustices, homicides, arrogances, ambitions, etcetera. (Cfr. 1, 28-32)

If we distinguish, according to the theological terminology, between the original originating sin (which according to the Yahwist does not exist) and the original originated sin (which would be the effect of the former), what Paul says is incoherent. According to him, the original originated sin consists in the sins that all men actually commit. It follows that no one can understand what Paul says without acknowledging that every man commits sin by himself. A thing of which God would be the only guilty one cannot be attributed to us. The truth is —as Hegel neatly understood in the Yahwist narrative— that "the natural man is egoist […] the naturality of the will is the egoism of the will" (PR III 115s).

"The essential content is that evil as such has its foundation in the spirit, neither in an action that happened once nor in an external naturality common to everybody" (EGP 289).

"The content is this: by nature man is not what he ought to be. He must be spirit, but the natural being is not spirit" (PR III 106).

"The evil is no other thing that the deepen-in-itself of the natural being of the spirit" (PG 539).

"To the extent that man wants the natural, this not purely natural but the negative against good" (Rph 139Z).

"Therefore, man is evil both by himself or by his nature, as well as through his reflection itself " (Rph 139A).

> What the empty fantasy imagines, namely, that the first condition of man was the state of innocence, is the state of naturality, and of animality. […] Innocence means not to have will. True, one is not evil in this state, but because of that one is not good either. The natural things, the animals, are all good; but the way they are good does not correspond to man. Man must be good with his will (PR III 115).

Because of what we have said, it would be an atrocious superficiality to confuse the Yahwist theses of the voluntariness of being with the Pelagian heresy. What the Pelagians defended was the innocence of man. Here, on the other hand, what is affirmed is that "such state of innocence, such heavenly state, belongs to animals. Paradise is a park in which only animals can remain, not men" (WG 728).

"Innocent is, therefore, the inaction of being a stone, not even being a child" (PG 334).

As superficial as the other thesis is to believe that the natural evilness of man is irremediable on the grounds of the Yahvist narration; the whole rest of that narrative speaks of the cure, as one can already see in Genesis 12, 3; 18, 19. As for Hegel, the next parts of our present chapter deal exclusively with that remedy. Furthermore, the voluntariness of the human evilness is in a way a manner of saying that it is curable.

On the contrary, what makes this evilness incurable is not recognizing its existence. The romanticism which prevails today about children and the savage man consist in that systematic blindness:

> According to that opinion, what makes man being what he ought not to be can have only be produced due to external contingencies or due to his inability of not consummating his natural skills, that is to say, the lack of opportunity in the free development of them. That is the hollow opinion of the pedagogy of our time, which, on the one hand, feeds and produces conceit, and on the other hand, does not search thoroughly, does not scrutinize in the depth of man, and hence does not produce any depth whatsoever, but moves rather in empty circles of self-indulgence and decadence (PR III 103).

It is absolutely distressing to see how these hollow opinions continue to prevail in spite of the accuracy with which Hegel denounced them and their disastrous consequences. The future of mankind is at stake, and we cannot continue to irresponsibly caress romantic and groundless apriorisms about the naturality of man! Especially now that we know we come from the animals.

Curiously enough, twenty two years after the death of Rousseau, the French found in the forest of Aveyron an unequivocal specimen of the natural man, a knave who was fourteen years old and who had not been modified either by society or by culture. Unlike other 'wolf-children' about which many stupid and unverifiable things are sill said all around the globe, Victor —as his unsuccessful educator, the acclaimed scientist Jean-Marc Itard— was the object of systematic observation by many of the best naturalists of the world of those times, naturalists whose documented testimonies have come down to us and have been recently compiled with the rigor of modern scholarship by the American investigator Harlan Lane. Before going to what really matters to us, let us make, out of curiosity, a selective extract of the observations that were drawn those days:

> Man's debt to nurture proved heavy indeed, even for the most elementary sensory discriminations, reflexes, and drives: the boy was indifferent to temperature and rejected clothing even in the coldest weather; he would put his hand in a fire; his eyes did not fixate; he reached alike for painted objects, objects in relief, and the image of objects reflected in a mirror; he did not sneeze, even with snuff, nor did he weep; he did not respond to

> loud voices; he did not recognize edible food by sight, but by smell; he preferred uncooked food and had no taste for sweets or hard drink; he had no emotional ties, no sexual expression, no speech; he had a peculiar gait and would occasionally run on all fours (1979, 101); "…fetid odors had no disagreeable effect on him." (126)

Let us address the question that directly concerns us. These are the words of Itard: "…the emotional faculties, equally slow in emerging from their long torpor, are subordinated to a profound egoism" (ibid. 161).

Another acclaimed naturalist, J.J. Virey, who also studied Victor carefully in the year 1800, wrote the following lines:

> It is astonishing how thoroughly this one idea absorbs him completely; he is always looking for something to eat, and he eats a lot […] is indeed fat. We might say that his mind is in his stomach; it is his life center. […] I am embarrassed to find natural man such an egoist; but I must report matters as they appeared to me. (ibid. 39)

In 1800, Virey also wrote: "His caretaker has never seen him show any sign of pity." (ibid. 43)

> I wanted to know if this child of nature would be content with his share if I put him with another person and gave them each an equal proportion of the same food – if he would respect that of his neighbor as property not belonging to him. But nothing of the sort transpired (ibid. 43).

The same thing is observed by Itard on the same date: "…he loves no one; he is attached to no one; and if he shows some preference for his caretaker, it is an expression of need and not the sentiment of gratitude." (Ibid. 39)

A refutation of Rousseau appeared few years after he launched to the world his prejudice in regard to natural goodness. As Hegel says, the "natural man is an egoist" (PR III 115).

By the way, we should notice that, against the Yahwist teaching that man tends to evil since his childhood, someone will probably try to entangle the facts and quote Mathew: "Verily I say unto you, Except ye be converted, and become as little children, ye shall not enter into the kingdom of heaven" (Mathew, 18, 3). But that was not the original formulation of Christ. The original sentence is the one we find in Mark

"Whosoever shall not receive the kingdom of God as a little child, he shall not enter therein" (Mark 10, 15). The first text clearly denotes a posterior reflection of the community or of Mathew himself: in order to receive the Kingdom of Heavens as children we must become children. But that was not the idea. The idea was to *receive* the Kingdom as children receive things, that is to say, as the children from the villages received Jesus when he went there to announce the Kingdom: receptive, without prejudiced nor preformed ideas, capable of hearing something truly new, with time availability, ready to follow up, leaving all other tasks aside. The original text speaks of the way of receiving, not of the way of being of children. Every specialist in the synoptic gospels knows that Mathew and Luke had before their eyes the text of Mark; in this case Luke (18, 17) preserved intact the formulation of Mark; the one that changed it was Mathew, and unfortunately, he wiped out the principal verb: to receive.

Some Bible scholars deny the historical authenticity of such phrase, let alone the formulation of Mathew. But they all agree on something: namely, that it does not speak at all about the innocence of children. For example Eduard Schweizer: "It is not about their purity or impurity" (1967, 117). And Walter Grundmann says: "Jesus does not presuppose a state of innocence in children" (1968, 207). On his part, D. E. Nineham says: "The point of comparison is not so much the innocence and humility [or obedience] of children" (1964, 268). Furthermore, C. E. B. Cranfield even attacks that interpretation: "To think of any subjective qualities of children here is to turn faith into a work." (1966, 324) We could make this list much larger.

It is irresponsible to affirm the illusion about natural goodness, especially now that we know that men come from the animals. Unlike in Hegel's time, today it is necessary to make focus on the biological aspect of our subject.

It may be convenient to start discussing with a great modern champion of all natural will, a convinced denier of the original sin, and, to a certain point, a biological expert: Abraham H. Maslow. In his book about motivation and personality, he dedicated the ninth chapter to hold that destructivity is not instinctive in man. For that purpose, however, he argues that the zoological beasts, which are apparently the most aggressive ones, do not attack motivated by pleasure but only to obtain food or defend themselves. However, he unwillingly says the following:

> Some animals apparently kill for the sake of killing, and are aggressive for no observable external reason. A fox that enters a henhouse may kill more hens than it could eat, and the cat that plays with the mouse is proverbial. Stags and other ungulate animals at rutting will look for fights, sometimes even abandoning their mates to do so. In many animals, even the higher ones, onset of old age seems to make them more vicious for apparently constitutional reasons, and previously mild animals will attack without provocation. In various species killing is not for the sake of food alone. (1970, 118)

I am afraid that the facts that Maslow grants in this passage can only be explained if there exists a natural and instinctive violence in the animal kingdom. The baroque discussion Marlow sets out on in order to distinguish between different types of aggression is pointless. At the end of the day, what this says to us is that the instinct of domination is not evil —as if a difference existed between being killed by domination and being killed for the sake of doing so. I will quote the next passage extensively because the urgency of the matter does not admit any kind of literary or stylistic scruples:

> When the higher animals are studied, attacking is found to be correlated more and more with dominance. [...] The animal's place in dominance hierarchy is in part determined by his successful aggression, and his place in the hierarchy determines in turn how much food will he get, whether or not he will have a mate, and other biological satisfactions. Practically all the cruelty manifested in these animals occurs only when it is necessary to validate dominance status, or to make a revolution in dominance status. How true this is for other species, I am not sure. But I do suspect that the phenomenon of territoriality, of attack on strangers, of jealous protection of the females, of attack on the weak or sick, and other phenomena that are often explained by instinctive aggression or cruelty are very often found to have been motivated by dominance rather than by a specific motivation to aggression for its own sake, e.g., this aggression may be means behavior rather than end behaviour. (1970, 119)

What a consolation for the attacked ones!

The fact that Maslow wants to explain the phenomenon of attacking the weak, the hurt and the sick ones —a phenomenon particularly frequent among the superior animals, and which in certain human tribes is simply instinctive repulsion— by appealing to the status of the domination, when the weak and the sick ones are actually those who jeopardize it the least, only denotes the power of dogmas on the mind of the scholar. But what Maslow and many of his followers lack is the philosophical approach. Despite its valiant defense of the natural, the hereditary and the biological in man, Maslow recognizes that we inherit from the animals the tendency to kill-in-order-to-obtain-more-food, the instinct of —aggression-to-obtain-more-sex, the impulse of attacking in order to achieve 'other biological satisfactions'. Now, that is not something evil in the animals, since we saw (V 6) that for them the end is the species. In the human realm, however, each and every one of the individuals is an end in itself. Hegel gave us the key to understand this: "To the extent that man wants the natural, this is no longer the natural but the negative against the good" (Rph 139 Z).

The destruction of the weak and sick suppresses form the world such genetic source. The only ones that breed are the strong ones, the better specimens of the animal in question. This mechanism is fabulous in the animal kingdom. There the species is the end.

The Nazis would praise the highly selective value that the facts compiled by Maslow have for the improvement of the race. But in the human realm each individual has infinite dignity and cannot be treated as a means. Whoever takes seriously the commandment 'Thou shall not kill', will only see in the above mentioned fact a conclusive argument which supports the Hegelian thesis that the being of man consists in tearing out the naturality in us. Man is man only insofar he ceases to be natural. Only he for whom his neighbor is an end and not a means can be considered a man:

> That man is good by nature is a doctrine of late that has a modern sense; one considers 'good' the inclinations and predispositions so that man is not good insofar he coincides with his concept but only insofar he empirically is, that means to say, only insofar the negative does not intervene in his vitality and existential functionality (PR III 102s).

There are other examples which are more deeply rooted in the biological than those registered by Maslow. I take from biologist Neal Griffith Smith member of the Smithsonian Institute, the following piece of information, which is tremendously disturbing if we take into account that in most mammals polygamy is the dominant system of reproduction:

> In a number of polygynous (mating of one male with more than one female) and promiscuous species, adult females outnumber adult males, sometimes by a factor of five or more. It has been erroneously suggested that this sexual imbalance is the cause of the polygynous mating system, in which one male has several female partners. It has been demonstrated, however, in all polygynous species so far studied, that the ratio of males to females is 50-50 at the time of birth; in many cases, this ratio persists until the cessation of parental care. Therefore, it is the polygynous relationship that causes the imbalance, not vice versa: because sexual selection is the dominant factor in a polygamous and promiscuous species, it results in a grater mortality of males than of females. (EB 14, 686, 2)

The masculine instinct of aggression is something which man inherits only because of the fact that he descends from animals. In the animal kingdom, where the species is an end, this instinct is marvelous because it makes the species improve qualitatively, since only the fittest males survive the confrontation and reproduce themselves. But in the human realm this aggression against the weak is absolutely unacceptable, because *every* human being is an end and not a means. To be sure, if one questions the validity of judgments of good and evil, this whole discussion becomes superfluous; but he who affirms that man is good by nature, or that it is good that the human race improves, is accepting the validity of judgments of good and evil; he is accepting the validity of morals. Now, no moral judgment can be grounded or justified if one denies that the person —for the simple fact of being one— is an end and not a means.

By nature man tends to destroy his rivals. Consequently, man is evil by nature.

And let us not lose sight from the fact that the tendency of polygamy is deeply rooted in the biological. The mass of an ovule is infinitely bigger than that of the sperm; sometimes it is many million times bigger. Unlike the sperm, the ovule contains cytoplasm, which is a warehouse of nutrients, so to speak. The female organism spends more energy in producing its gamete than that which the male organism employs doing the same. The female tends to be very selective; it tends to mate by nature with the best male exemplaries of its species, for she cannot risk her fecundity of one year or of her entire lifespan. On the contrary, the male can allow himself a great number of bad choices, because his organism wears out very little by producing sperm. Polygamy is natural in the male, while the natural selectivity of the female favors the confrontation between males, for all the males have by definition a sexual instinct, not because they are the best exempes of the species.

Regardless of sex and reproduction, the instinctive aggressiveness is a *fact* which is corroborated at behavioral and physiological levels. Let us mention briefly the behavioral aspect, since the physiological one is much more impressive.

Zoologists distinguish between gregarious and solitary species. The termites seem to belong to the first group, and it has been thoroughly documented that they eventually end up eating each other (cf. EB 21, 612, 2). But indefectible aggressive behavior has been observed in solitary species when we gather many individuals in an enclosed space. When that happens, we can even observe cannibalism among both rats and termites (cf. EB 14, 687, 1). In the case of any solitary species, one only needs to gather in a closed space two individuals, and the result is either that one murders the other or that one becomes completely dominated.

In regard to the physiological one, neurophysiology has made a hideous discovery by means of encephalography: rage, resentment and anger are emotions *pleasant* for the organism, positive feelings, encouraging dispositions. Let us summarize the technical procedure that has provided us with that conclusion:

"Little animals could learn systematically to connect to or disconnect themselves from an electric stream by pushing some pedals connected to their hypothalamus, the inner part of the diencephalon. The intensity of the current has unequivocally something to do with pleasantness and unpleasantness, for when they are subjected to a soft stimulation they immediately learn to get away from the pedal that disconnects them. Now, since one is making of them an encephalogram, the experimenter can distinguish two different types of waves in the movements of pleasure and displeasure; the pleasant ones are wide and large wages; the negative ones are narrow and short" (Cfr. Grastyán: EB 18, 354s).

Once he has come to realize that, the scientist can place the animals in vital, real situations and observe in the screen which kind of waves are produced by different situations. When there is a feeling of fear or anxiety, the waves indicate unpleasantness and suffering; however, when the animal attacks he becomes angry, the waves turn slow and wide, which are characteristics of contempt and self-reinforcement. This is the horrendous discovery we mentioned before.

Not that this content is entirely new, but anybody that has certain honesty of analysis knows that there is pleasure in the act of carrying out an act of revenge and that deeds performed with rage are self-reinforcing. The first thing to say is that the above mentioned experiments cannot be denied or neglected. Second, and more important, they demonstrate that our in itself pleasant, gratuitous aggressiveness is not the product of culture or education, those universal villains to which romanticism attributes all possible evil in order to remain with the reveries of the natural goodness of man. No. Cruelty is something inherited from the animals, it is a *natural* element of man; and man becomes man insofar and to the degree that he abandons naturality. Man is an animal while he remains natural. Insofar he is a man, man has no nature.

To top it all, scientists have been able to discern between the vascular and hormonal changes that come along with pleasure and the vascular and hormonal changes that come along suffering and anxiety: one has found the presence of the first in the moments of rage and resentment. This kind of reaction have confirmed this: fury and aggression unconsciously trigger typical movements towards the object, while the unpleasant experience triggers typical movements in which one grows apart.

To summarize what we have said, let us repeat with Hegel: "The evil is no other thing that the deepen-in-itself of the natural being of the spirit" (PG 539).

In order to go further with our exposition, we need to detonate once and for all the Hegelian bomb: "Everything that man is, he owes it to the State; only there he has his essence" (VG 111).

The State, as we will demonstrate later on, consists *only* in the set of rights and duties which bind man. Now, we have seen (III, 7) that self-consciousness can only be produced by the ethical demands that others address to me and which are called duties; but that which distinguishes man from animals, that which makes him truly a man, is self-awareness. This why Hegel says that everything that man is he owes it to the State. Hegel was the first one to understand in the modernity that the Aristotelian expression *zoon politikón* is the *definition* of man. What has hindered the most that one understands Hegel is the chimerical belief that man is good by nature. Only the State, only a set of rights and duties, makes him good; for only the State pulls him off from animality.

In a like way, we showed that (V, 1), if man follows nature, he is not free. When the path of behavior is not determined by the self but by an arbitrary impulse, the kind of action we are dealing with is heteronomous: no one can speak there of freedom. It follows that the set of duties and rights is what makes man free. With the first ethical answer man begins to be free; he begins to be good and, consequently, he begins to be a man. "The natural want [...] is worse that the bad want" (Rph, Notiz zu 139 A). "What is morally evil is the worst; it that which absolutely ought not to be; however, nature is even worse" (ibid.).

Even the most primitive State or *polis* contains the roles and expectation which make man a man by granting him responsibility —not killing, honesty, respect for the others' existence, etcetera. This is what is essential to the State: a set of duties and rights.

"One renounces to the particularity of will in those uncouth statal situations" (VG 113).

"To be sure, one can verify the existence of many situations of savageness, but [...] despite how ignorant these situations are, they are bound to institutions which, as one says, limit freedom" (VG 116).

"The substantial of ethical relationships [...] is already present in an uncultivated society" (PR I 158).

In this context —as we have said before— the most unlikely ally we can count on is Rousseau himself: "Good social institutions are those that know best how to denature man" (*Emile*, I).

Man is not free by nature: this Hegelian thesis, which only makes explicit what we have demonstrated (V, 1) is undoubtedly one of the most important truths both in political philosophy and anthropology. Neither liberals nor leftists seem capable to understand it. The truth is that neither liberalism nor communism can define freedom.

If man is *not* free by nature, then saying that duties 'limit' freedom is absurd. That is the context of Hegels phrases (VG 116), as they say in German *sogenannt*.

"Duty is not the limitation of freedom, but only the abstraction of it, that means to say, the lack of freedom (Rph 149 Z).

"In duty man frees himself for true freedom" (Rph 149).

"As ethicity, *true* freedom consists in that the will does not have as an end, subjective and selfish contents, but rather universal content (EPW 469 A).

"Merely natural will is in itself violence against the idea of freedom in itself, which one must protect against uncultivated will" (Rph 93).

"As a matter of fact, every law that is truly a law is a freedom, since it contains a rational determination of the objective spirit and hence the content of freedom. On the contrary, it is now very common to say that everyone must limit their freedom in relation to the freedom of others, and that the State is that situation of reciprocal limitation, and that the laws are limitations. In such imaginations one conceives truth as a caprice or a whim" (EPW 539 A).

> But the ambiguity of Rousseau begins soon. Man is free, that is the substantial nature of man; and it is not only suppressed by the State for the first time, but in fact is constituted for the very first time within it. The freedom of nature, the aptitude for freedom, is not true freedom; for only the State is the realization of freedom (GP III 307).

"In reality this limitation is the condition of liberation; society and State are rather those situations in which freedom becomes true" (VG 118).

As we said before (v, 1), every mistake in this regard has its origin in believing that freedom is something negative, that is to say, lack of something. It is very easy for the natural man to possess from the beginning something which consists in nothing. Aside from the fact that such conception does not define truth and hence does not know what it is speaking about, the absurd that immediately follows from it is that stones would also be free. Natural goodness has a similar origin: if goodness consisted in the lack of something, it follows then that man has it by nature, for one does not need to put many endeavors in obtaining nothingness. That is the origin of Rousseau's apocalyptic craziness: he understands the natural man as a solitary animal, and hence he does not harm anybody, for the sole and simple reason that there is no one around him, and in that sense this man is good: a goodness that consists in a *negation* or in nothingness. Obviously, Rousseau never understood that intersubjectivity is the only thing which can make a man *exist* and that a 'solitary man' is a *contradictio in terminis*. Besides, with that criterion, even rocks would be morally good.

"In the State freedom is realized *positively* and is in itself objective. However, this should not be understood as if the subjective will of the subject came to its realization and enjoyment *by means* of the general will, in other words, as if the former was a means for the latter. The State is not a coming together of man in which the freedom of all individuals ought to be limited. We would be conceiving freedom as something merely negative if we imagine that each subject limits its freedom with other subjects, so that this common limitation, this common 'getting in the way' of the others, left everyone with a place in which he could let go. As a matter of fact, only Law, ethicity and the State are the *positive realization* and achievement of freedom. To be sure, the arbitrariness of the individuals *is not* freedom. What is limited is a freedom that consists of caprice and that has nothing to do with particularities and necessities" (VG 111). (The italics are mine).

What the State and duty must limit is animality and "deepen-in-itself of the natural being", which is not freedom but naturality. The conception we have criticized would need to affirm the absurdity that loving the neighbor is a limitation of oneself, when in fact loving the neighbor is to identify oneself with the others. Nature is not freedom. Freedom is something extremely positive which implies breaking out from the natural, for no natural thing is free.

2. Happiness?

It is amazing that neither Philosophy nor Theology have noticed that the thesis which states that the end of man is happiness is a huge immorality. If this thesis was correct, all my neighbors would be only a means and nothing else in order to obtain my final happiness —God would be a mere means for that purpose as well. It seems almost impossible that such a terribly perverse thesis, in spite of being denounced by Hegel with perfect clarity more than a century and a half ago, is nowadays held —implicitly or explicitly— in politics and even in Theology.

As a means of introduction, let us address the relation between this thesis and our previous theme.

It is obvious that the romanticism of men of letters and poets, in order to make beautiful their bucolic descriptions, had to suppose *a priori* that the natural man was happy. It is what Hegel calls "the frivolous dream of natural happiness" (EPW 475 A).

This apriorism is evident in Rousseau: "Every man wants to be happy, but in order to become happy he must begin by knowing what happiness is. The happiness of natural man is as simple as his life: it consists in the absence of pain. Health, freedom, the necessaries of life are its elements" (*Emile*, III).

Just like goodness and freedom consist in a *not* — in a lack and absence of something—, so happiness consists in a 'not suffering'. A thing that does not consist of anything is something easy for the natural man to possess from the very beginning.

By idealizing the natural life, romanticism did not realize that those who are civilized are the ones who *positively* enjoy the fields, the woods, the mountains and the sea. The beauty of it all would be nothing but a mystery to the 'natural man'. Moreover, like Norbert Elias has pointed out, "for the primitive men the natural space is to a larger degree a danger zone; it is full of perils that the civilized man does not know any more of" (1977 II, 405). In order to enjoy natural beauties, one demands the pacification of the environment and the affectivity which are introduced by civilization; one demands that forests and fields are no longer the place where men and animals hunt each other. But still today, in our pacified countries, it is by no means true that the uncultivated man is more capable of enjoying the natural beauties than the cultivated one. Hegel says:

"Granted: the savage man does not know any huge amount of pain and unhappiness; that is, something merely negative; freedom, however, must be essentially affirmative. Only the benefits of supreme consciousness are the benefits of the affirmative freedom." (WG 775)

The observation that the English novelist Willkie Collins made in the middle of the nineteenth century is extremely accurate:

> Admiration of those beauties of the inanimate world, which modern poetry so largely and so eloquently describes, is not, even in the best of us, one of the original instincts of our nature. As children, we none of us possess it. No uninstructed man or woman possesses it. Those whose lives are most exclusively passed amid the ever-changing wonders of sea and land are also those who are most universally insensible to every aspect of Nature not directly associated with the human interest of their calling. Our capacity of appreciating the beauties of the earth we live on is, in truth, one of the civilized accomplishments which we all learn as an Art; and, more, that very capacity is rarely practiced by any of us except when our minds are most indolent and most unoccupied (The Woman in White I, viii).

Without referring to the enjoyment of natural beauties, the Cyrenaics of the fourth and fifth centuries B.C. —a philosophical school whose strongest point of focus was happiness— stated that only culture and reflection make man capable for joy. Hegel summarizes Aristippus's philosophy thus: the principle of fruition "embraces the feature that culture of spirit and thought is an ineludible condition to achieve delight" (GP I 541).

Natural happiness seems to be something impossible. It is with all certainty a dogmatic and superficial apriorism.

Because it is clear that happiness cannot be defined as something negative, like a lack of something, e.g. like absence of suffering; for then, the stones would be happy.

If what they mean to say is that absence of suffering is the mere condition for happiness, one could agree with that, but it is obvious that we still need a definition of happiness.

In spite of what Rousseau says, health and minimal material goods are evidently not enough, because there are people who are perfectly healthy and have all the things in the world and yet are tremendously unhappy.

Perhaps Rousseau would say that those are mere *ideas*. But one could reply the following: first, that happiness and unhappiness are precisely ideas, and hence health and minimal material goods would become, again, a simple condition and we would still lack a definition of happiness. Second, if unhappiness is a mere idea, Rousseau would be obviously supposing that the natural man is happy because he does not have ideas. He holds a conception of happiness which consists in the mere absence of something. A tree that is healthy and does not need anything would have to be considered happy. In short, the Rousseaunian definition is untenable.

Fortunately, Rousseau adds after the above quoted text: "Another thing is the happiness of the moral man, but we will not speak about it here". More fortunately, many years afterwards and having changed his mind in his *Political Fragments,* Rousseau was finally able to recognize this: "But the meaning of the term happiness, which is much undetermined among individuals, is even to a greater extent so among the nations" (1964. 509).

This is precisely the problem: men of letters and politicians —and even theologians and philosophers— have induced mankind to chase eagerly an ideal which nobody knows anything from.

"Happiness is the only imaginary and abstract universality of a content which simply *must* be" (EPW. 480).

"...the whim which in happiness gives or not himself a goal" (ibid.).

It is an imaginary configuration whose only content is the unreachable. It seems to be a goal, but since it lacks content it isn't one at all.

Regardless of the harshness of the example, we could say that the whole situation is like putting a carrot in the end of some stick so that some animal chases it. Except for the fact that the carrot is something concrete, while happiness does not have content. It is, by definition, the *processus in indefinitum*.

It would be very simple to say that happiness is the satisfaction of impulses. But, in the first place, some impulses contradict each other. Thus the satisfaction of one inclination is the dissatisfaction of another. It would not be enough to satisfy only one of them, for the distinction between happy and unhappy men would disappear and the concept in question would lose all cognitive value, since man are always satisfying some impulse, despite how mundane it may be.

In the second place, there are killer impulses too, or at least, inclinations that are harmful to the others; therefore, one cannot say that the end of man is the satisfaction of impulses. In this context, there is no difference here between natural and acquired impulses; the fact is that they are there. If one answers that one must distinguish between different kind of impulses, then the definition we are dealing with proves to be inefficient, for all impulse demands satisfaction, and the criterion to discern between impulses would be based in the fact that they are impulses; one needs as a higher criterion and a new content that are not provided by this definition. Therefore, the previous definition remains undetermined. Besides, let us not forget that there are impulses which are harmful to the subject and hence the satisfaction of them cannot be the definition of happiness.

In the third place, experimental psychology nowadays has demonstrated that our most decisive impulses are *acquired*; they have their origin in education, social influence and culture. Now, this makes of the expression 'satisfaction of impulses' a completely undetermined term, because education and social influence can create all kind of impulses, which can be contradictory and incompatible. But if the term in question does not have any determined content, it does not work as a definition. In order to reinforce this third point we will discuss in short the thesis of psychologist Judson B. Brown.

The impression that the definition we have criticized leaves is that we need to satisfy only those impulses which aim at happiness. But if that is so, then the tautology and the lack of content therein become evident, for it turns out that happiness is the satisfaction of the impulse towards happiness. There can be no more lack of content than in an expression which only in appearance defines things. Hegel was right: "happiness is the undetermined" (PR II, II 228).

Let us now present an historic argument of appreciable forcefulness about the intrinsic indetermination of eudemonism: Epicurus, the most acclaimed specialist in the pursuit of the pleasant, arrives to the same 'ideal' of behavior that Stoicism, whose asceticism and contempt against the pleasant are paradigmatic. This fact is not fortuitous: who talks about happiness imagines something 'pleasant', but precisely the 'pleasant' lacks content and is undetermined: "one can understand by 'pleasant' anything" (GP I 539). The result is that "in Epicurus, the wise man is described with the same characteristics, negatives by the way, that in the stoics" (GP II 325). "The systems of that time, Stoicism, Epicureism and skepticism, although opposed to each other, lead at the end of the day to the same, namely, to make the spirit indifferent with respect to everything that reality has to offer" (WG 718).

The logical process of Epicurus is not fortuitous. It places from the very beginning pleasure and contentment as the end of men; but it realistically understands two things that make of that very criterion something thwarted and equivocal. First, not only the corporeal produces pleasure, but culture and intelligence too, as Aristippus affirmed. Furthermore, culture and intelligence provide more pleasure than the corporeal and are a condition *sine qua non* of true joy. Second: the negative, the absence of pain and nuisance, must be considered as an element of happiness it and may be the true constitutive of it. The pursuit of pleasure is not a univocal compass that we could follow to the end, because that pursuit is always interrupted by the avoidance of the unpleasant.

And the worst is that, when both considerations are carried out, the criterion falls apart completely, because there are displeasures of the soul —e.g. fear, anxiety and concern —that probably hinder one more than the displeasures of the body. If we add remorse —which necessarily must be added, because it is a psychological fact, despite what the immoralist says about it is objective validity—, the eudemonist criterion becomes something ludicrous, because when morals steps in joy has ceased to be the norm. The hedonistic logic, the consequent pursuit of happiness, is what makes that Epicurus prefer tranquility, the *ataraxia*, which is very similar to nirvana and nothingness. What started as an easy pursuit of pleasure ends up being a rigid discipline of affections and passions that preserves imperturbability. One could not come up with a better demonstration of the vacuity of the pseudoconcept of happiness.

By other way, the cyrenaic Hegesias arrived at that negative and ataraxic result before Epicurus did. He proceeded thus: It is necessary to choose between complete situations, not between an aspect of a situation and an aspect of other situation; the real is not the pleasure that a man feels in a given moment, but all what that man feels in that moment, because it is possible that he experiences displeasure with regard to some other thing. Now, a completely happy situation does not exist. How could we know if the welfare of a dead man with no anxieties or troubles is superior to that of a living man who experiences some sort of pleasures? Why do we prefer to be alive? The disciples of Hegesias started to commit suicide and the authorities were forced to forbid him to teach.

Hegel comments this very well:

> The body, says Hegesias, is distressed by several ailments and the soul suffers with it; this is why it is indifferent to choose between life and death. As such, Hegesias says, nothing is pleasant or unpleasant; it is futile to proclaim the liking as the normative; because, on the contrary, it is the null what does not have any determination in itself; it is negation of objective determination (GP I 548).

The refutation of eudemonism is not only the question: what is that thing called happiness? There is a question that goes even deeper: who said that happiness is what man wants the most?

As Victor Hugo very well remarked: "The reflective spirits use this expression very little: the happy and the unhappy ones" (*Les Miserables* IV, VII, i). In this regard, we find a very eloquent of Rousseau in his *Political Fragments*: "Hurry to abandon the laws, for they only are useful to make you happy" (1964, 556).

And in the *Emile* we read: "They say we are indifferent to everything but self-interest; yet we find our consolation in our sufferings in the charms of friendship and humanity, and even in our pleasures we should be too lonely and miserable if we had no one to share them with us." (Book IV)

"One says that one contributes to the common good by one´s own interest. Where does the idea come from according to which the just contributes against his own interest? How can one die looking for his own benefit?" (*ibid*).

There is a very widespread dogmatism —whose purpose is, undoubtedly, to justify one's own selfishness— that says that in the last instance the only thing that motivates man is self-interest. First, this dogma does not say anything, because it cannot provide the expression 'self-interest' with a meaning whatsoever. Second, and more important, this dogmatism refutes itself when it is asked: how can one die in self-interest? How can one find satisfaction with acts which are completely disinterested? They can only juggle with words when they try to answer that. They end up recognizing, implicitly or explicitly, that man is capable of acting without pursuing his own interest. To say that one takes benefit from not taking benefit is only a game of words which does not respect the principle of contradiction.

In order to talk about of 'impulse' towards one's own benefit, the mentioned dogma needs to focus on introspection, because no impulse can be called an empirical data, not even hungriness. But if we rely on introspection, we find there other motives besides self-interest, which is enough to refute this dogma, for according to it the only thing that moves us is self-interest.

Morals and my neighbors are ends in themselves. Those moralists who believe that morals are a means to obtain the final happiness have not read Kant, let alone Hegel. They have not come to realize the difference between a categorical and a hypothetical imperative. And they do not know anything about God.

About the human plenitude involved in the truly moral act Hegel makes the most important and precise observation: "This happiness, in contrast with the other one, could be called real happiness, but in that very moment happiness becomes an inadequate expression [...] it is a reality that becomes deformed if one calls it happiness" (GP II 288s).

"To direct oneself towards happiness and spiritual joy, to chitchat about the wonders and delights of science and art, is something futile; for the very thing that occupies oneself there does not have the form of pleasure; in other words, that entire conception is suppressed" (GP II 289).

"All that empty speech is left behind and loses all interest. The true spirit consists in dealing with the thing in itself, with something that is in itself universal and not a means for pleasure, that is to say, not as if it were the constant reflection of the relation with oneself as individual" (GP II 289).

The last three texts are the key to the whole issue. It is completely false that in the genuine moral act —and even the activities that are not as elevated as that one, such as the scientific quests and the creation of art— one is relating to oneself as individual; the pleasure is experience of the individual as such. About the true morals we will talk later.

The above mentioned dogmatism does not need to verify if in a very specific case the motive was self-interest or not; that dogmatism believes to know everything and does not need to verify what it says. It is one of those pampered theses which need no contact whatsoever with reality and, consequently, cannot refute anything at all. Whether things are A or not-A is indifferent to them.

Another example would be the thesis that the only motive of man is the desire of self-sacrifice or of mortifying himself. In light of the bacchanals and *bons vivants*, it would be enough to say that men adopt such conducts even going against their own liking and inclinations, and they pretend to feel joy so that the triumph over themselves becomes of a more sophisticated nature. As in the case we are dealing with, this second apriorism acts stubbornly despite what reality may say. It is simply a mental toy, not a real knowledge.

In order to refute the second apriorism above mentioned, the eudemonists would need to use the introspection, a recourse in which we would certainly find some hedonistic impulses. But they are lost; the introspection also refutes them, because it testifies that there can be in men other motives different than self-interest; for instance, the presence of the moral imperative.

> To be sure, there is satisfaction in universal history, but that is not what is called happiness, for it is the happiness that goes beyond particular interests. The ends that have real value in universal history need to be kept firmly by means of strong will and energy. The relevant individuals in universal history —who pursuit such ends— were rewarded, it is true, but to be happy was not what they wanted (VG 92s).

> Delight is something secondary, concomitant to the fact. When the substantial is realized, delight is added to the extent in which the work is perceived to be the work of the subject. Who sets out on the quest of delight, only looks for oneself in the accidental. Who cares of great works and interests, only looks for the realization of the thing in itself. *He directs himself towards the substantial, does not remember himself in it, and forgets about himself in the thing*. People use to pity men of great interests and works, because they have little satisfactions, that is to say, because they live only in the thing, not in the accidental (NH 255).

One should understand here clearly that I am not encouraging asceticism or austerity. I am only pointing out that happiness —besides being something completely undetermined and with no definable content— is not the only or the deepest motive of man.

"As an individual one must strive for one's reasonable wellbeing and the more pleasant the better. But one does not need to make big fuzz or great speeches about this, as if this were a matter of great importance and intelligence" (GP II 289).

Hegel warns us explicitly about Stoicism:

> But to reduce as much as possible our dependence from necessities is a mere abstract freedom. Once we have achieved indifference with respect to necessities, true freedom consists not in the diminution of them but in being free in joy itself, and in remaining with ethicity in being part of the life of men in law. On the other hand, the abstract freedom suppresses the ethicity: the individual retracts himself to his subjectivity; such freedom, hence, is a factor of immorality (GP I 553).

Once we have made that clear, we can move forward. The eudemonist apriorism has become so unreflectively widespread, that theologians have posed as the final goal and end of man the beatific vision of God. Beatific meaning: what makes one happy.

Being God a spiritual entity, to talk about 'seeing him' is nothing more than a metaphor, and in such a serious issue like this one there is no room for literature. To bring up the words of Saint Paul, who says that we will see God "face to face" (1 Corinthians 13, 12), is something that a careful treatise writer should not do, knowing that God does not have a body. Even Paul in his first letter to Timothy explicitly warns: "whom nobody has seen and cannot be seen" (1 Timothy 6, 6).

The same warns Saint John, making thematically evident that God can only be known by means of loving the neighbor: "No one has ever seen God, but as long as we love one another, God remains in us and his love comes to its perfection in us" (1 John 4, 12). Blass and DeBrunner, two of the most intelligent scholars of the New Testament, point out that those forms of aorist (like our 'has seen') does not refer to the past but are in fact gnomic aorists; the expression we are dealing with means 'has never been seen', in an intemporal form. And when the passage adds that our love for God has reached its perfection if we love the neighbor, which evidently means there is no knowledge of God that goes beyond this.

The postulate of a 'vision' that we criticized suffers from another evident superficiality: it believes that the best way to know a person is looking at it as if it were an object. As a matter of fact, to know a person means to be addressed by another: the essence itself of a person is dialogical and intersubjective. The character and depth of his appeal is what matters; his unconformities and his conformities, his ambitions and his plans, his wishes and his preferences, his patience and his impatience, his historical burden, his lucidity about reality, his degree of consciousness in respect to the world and other people, his tolerance, his trajectory, his dialogical intensity, his capacity or incapacity to sacrifice, and above all, his degree of true morality, his affectivity, his intelligence, his keeping his word, his honesty, his delicacy, his tenderness. The spirit is act, it consists in its acts, especially God, who the Scholastic philosophers called 'pure act'. To imagine a 'divine essence' which does not consist in acts is tantamount to imagine God as a material being.

And let us now address the central point. Who said that knowing God deeply has to be 'delightful' and 'beatific'? We need to say with Hegel: "it is a reality that becomes deformed if one calls it happiness" (GP II 289). Pleasure and pleasant are words completely inadequate and even childish, when they mean to describe an overwhelming astonishment; an amazement that cannot decrease or increase, whose nature inspires piercing veneration. It is a boundless wonder which never ceases to be disturbing and powerful. T he predominance of the eudemonist terminology confused theologians.

As we have seen, Hegel defends eagerly and coherently the immortality of the individual soul. However, that does not mean at all 'another world'. For a long time, theologians were deceived by the expression 'the Kingdom of Heaven', which, as the parallel passages of Mark and Luke can bear witness, was an expression introduced by Mathew instead of the original expression of Christ, the 'Kingdom of God'. All the exegetists —whether liberals or conservatives— know that Mathew uses there the circumlocution employed in later Judaism in order to avoid pronouncing the name of God, which was a poorly respect towards the divinity and a merely legalist interpretation of the prohibition of using the name of God in vain. Mathew himself teaches us explicitly in the parable of the tares that the place where the Kingdom is reestablished is no other "the world" (13, 38), and that Christ will return to such Kindgom to expel from it all the people who perform inequity (13, 41). And in the teaching that our Savior taught us, we do not read "take us to thy Kingdom" but rather "thy Kingdom come" (6, 10).

The resurrection of the bodies is not only an argument against what we have been saying. It reinforces our entire point. The bodies need a physical ground; they have to be in this world. There is no single Biblical text that says that the Kingdom of God belongs to another world.

Now, the Kindgom which Christ started to conform was a *State*. As Hegel says: "the reality of the Kingdom of Heavens is the State" (FR 231). We will see later on how little they understand the *Philosophy of Right* those who do not see that the reality called State is a part of universal history (Rph. 341, 360) and that the universal history is directed to the realization of the Kingdom of God.

One cannot put into question that the Kingdom that Christ wanted to gather and form —and of which he always said 'it has come' (Mathew 12, 28; Luke 11, 20), the verb *phtháno* does not mean 'to approach' but 'to come' — consists in the set of persons bounded by duties and rights. And in the parable of the grain of mustard seed (Mark 4, 30-32) the thesis is that this Kingdom, although it may now be of small proportions, will grow one day more than any other.

This kingdom is the only end of man. Only in this kingdom, true ethicity among mankind is reached and God becomes truly conceivable. It is intolerable the immorality that says that my neighbors and my interrelation with them are a means to the consecution of my heavenly happiness or to the consecution of any other thing. Man must find his own realization and plenitude in responding positively and creatively to this continuous ethical appeal, to this intersubjectivity in which the kingdom of God consists. As Hegel says, the subject 'must find his satisfaction in his ethical situation' (VG 264).

This is the point where Kant failed in a lamentable way, and this is the reason why Hegel criticizes the Kantian morals, in spite of the fact that its distinction of categorical and hypothetical imperatives is an undisputable merit of Kant and the key to overcome definitively eudemonism. The Kantian postulate of a future harmony between the moral good and the physical good (= happiness) hinders that the end of man is morality and the neighbor. "The harmony is not present, it is not real; it only must be. The postulate itself is perennizing" (GP III 370s). "The immorality expresses here precisely what it is: that morality itself is not what matters, but happiness as such, with any relation to the former" (PG 440).

With regard to Kant, Jacobi and Fichte, Hegel says: "It is false that these philosophies abandon eudemonism; they rather perfect it to its maximum extent" (JS 294).

As I have said, Hegel, makes this objectively irrefutable criticism, in spite of the fact that he recognizes that the key to overcome eudemonism is in Kant: "before the Kantian philosophy, morals, as eudemonism, was grounded on the notion of happiness" (GP I 186).

Intersubjectivity —i.e. the vinculation and unification among men— is not a means for something else, but rather an end in itself. This is the key to Hegel's *Philosophy of Right*, the most misunderstood work within his oeuvre: "The unification as such is in itself the true content and end, and the destiny of individuals is to realize a universal life" (Rph 258A).

"We now know that the ethical and the just in the State is also the divine and the commandment of God, and that insofar the content there is no other thing which is higher or holier" (WG 888).

"The divine is no longer represented as a beyond" (WG). Those who questioned that man could find his own realization in intersubjectivity and perfect plenitude understood him wrongly and ignore the most recent physiological and anthropological discoveries. Psychologist Abraham Maslow —an impartial witness in this case— says this very clearly: "At the highest level of life [...] duty *is* pleasure" (1970, 102).

Those who put this into question commit two mistakes. Fist, they still have the immoral belief that one needs to satisfy his natural and animal impulses, as if we did not possess also killer impulses that enter into conflict with the other impulses. Second, and most important, they have not realized that only the *acquired* impulses are in fact the only ones that count for experimental psychology nowadays, because they obliterate the natural ones and erase them practically from the list of motivations which demand to be satiated. Psychologist Judson B. Brown says:

> ...the doctrines which hold instinctive sources of motivation to be of significance for human behavior have all but vanished from the psychological scene. These and other influences have led many to the belief that the important human motives are produced by learning during the processes of socialization and acculturation. (Teevan/Birney, 1964, 80)

If the word happiness had a meaning, it could only consist in the satisfaction and fulfillment of the *acquired* impulses. The Hegelian thesis according to which the subject "must find his satisfaction in his ethical situation" (VG 264) not only is it not unreal, but it demonstrates how Philosophy anticipated the discoveries that Psychology and Anthropology have only been able to make more recently.

"Delight is undetermined in respect to its content, because we can find it in every object" (NH 256).

"The individual must find somehow, in the fulfillment of his duty, his own interest, his own satisfaction, what matters to him" (Rph 261 A).

"In its concept, *passion* contains the fact of finding itself in a particularity of the determination of the will, in which all the subjectivity of man is rooted, despite what the content of such determination may be. Due to this formal reason, passion is neither good or evil; such form only express that the subject has put in its content the whole vital interest of his spirit, of his talent, of his character, of his joy" (EPW 474A).

All the doubts and skepticisms as whether man can find his own fulfillment and plenitude in his ethical intersubjective relations have their practical origin in a lack of culture, a lack of experience of true culture; and in the theoretical level, their cause is the traditional and false belief in the real distinction of the faculties, that is to say, to believe that the understanding, the will and the emotions are completely different to each other.

> For the abstract intellect the difficulty consists in freeing oneself from the separation which he arbitrarily introduced between the faculties of the soul, between the sentiment and the thinking spirit, and in coming to the idea that *there is only one* reason in man, in feeling, willing and thinking. God, the Law, ethicity, can *be felt*. But the feeling is no other thing that the form of the immediate and characteristic individuality of the subject, in which these contents can be posited, as well as any other objective content which is granted validity by consciousness (EPW 471A).

But let us repeat what we have said: the fundamental mistake of eudemonism, besides its intrinsic immorality, is to believe that what man wants the most is happiness; that man *looks for* his own satisfaction and happiness in everything he does. In fact, things are not like that: "delight is something secondary, concomitant to the fact" (NH 255), it is by no means an end.

For marriage, it is time now to denounce the superficiality of all the theories that do not recognize this: for the married couple the other is an end in his/herself. Those who ignore this, ignore the true nature of love.

"Precisely, marriage has to be beyond happiness and unhappiness" (Rph, Notiz zu 162).

"Marriage is essentially an ethical relationship" (Rph 161 Z).

The spouse is an end; she is not a means for my happiness. Having the other as an end brings one a depth and a plenitude which are far beyond that thing called happiness.

In the summer of 1811, Hegel wrote this to his future wife: "marriage is essentially a religious bond; love needs to be completed with a superior element of what it is for and of itself" (Kaufmann 1966, 333). The other cannot be seen as an end if he or she is not looked upon as identified with God Himself; the agnostic humanism falls short with its own epistemology and the reflection over its own act.

Hegel distinguishes consistently between *beseligend* and *beglückend*. In English, it is convenient to distinguish between joy and happiness. Joy is not something one pursuits. Happiness, on the other hand, is pursued: happiness is by definition a goal, something one goes after. Automatically, morality is destroyed with that criterion, for the neighbor becomes a means.

In another letter of that summer, Hegel tells her future wife:

> I have hurt you by giving the impression that I condemn —as if they were your own principles of thought and behavior— moral conceptions which I am bound to condemn. I will only tell you two things. On the one hand, I condemn such conceptions because they suppress the difference between what the heart likes and duty, or in other words, they eliminate the latter and destroy morals. On the other hand, and this is what really matters, I beg you not to think of me as somebody that ascribes those conceptions to you with their belonging consequences. I regard them only insofar they are present in your reflection, but not as if you were fully aware of them and their consequences (ibid.).

One should not believe that the rejection of eudemonism is something that only concerns the subtleties and the wise men. It concerns the very core of morality itself. Hegel accused the Kantian, eudemonist postulate of a future harmony for being immoral and nothing else. The eudemonist transforms the neighbor into a means, and that is immorality without adjectives or subtleties of any kind.

3. Ethicity

Now we can address the message contained in the structure itself of the *Philosophy of Right,* whose understanding depends entirely on the meaning of the word ethicity (*Sittlichkeit*) —a term that was deliberately chosen by Hegel in order to get over the Kantian inmoral morality (*Moralität*) once and for all. It is indeed true that eudemonism is not the only thing that Hegel disapproves of about Kant's position; however, the criticism that we have seen shows that Hegel assumes in its integrity, and even with more exigency, the Kantian distinction between categorical and conditional imperatives; the categorical imperative is an end in itself, not a means for something else.

Previously, I have employed the terms 'ethical' and 'moral' indifferently. In what follows I will not do that anymore. I beg the reader to have this into account. "Morality and ethicity, words that are ordinarily taken as synonyms, are taken here with essentially different meanings" (Rph 33). The etymology of both words is the same; the root of *both* is: custom. But "since both different words do exist, this does not prevent that one uses them for distinct concepts" (ibid.).

After what has been said about happiness, one cannot assume that the distinction introduced by Hegel is a subtlety of nomenclature. The reader can be sure of this: it is the most important *content* in moral, political and juridical philosophy. In this case, in order to understand the terms, we have to do without the etymology, since they mean in that regard exactly the same thing. As we shall see, what ethicity means is: the only true morality. Hegel shows that the morality of the treatise writers, not only of Kant, is sheer immorality: when they distinguish between morals and right, neither morals nor right ascribe legitimately those terms to themselves, since by its own concept "right is not something sacred at all" (Rph 30). Hegel refutes not only the moralists but also the jurists, more particularly, juridical positivism, which by no means is an invention that Hans Kelsen made in our century.

Hegel makes clear that the first part of his work —which is devoted to right without ethicity— has as its object of study abstract right, the *false* right. Bourgeois commentators have not wanted to take charge of that insisting warning, because there, in false right, is where the right of property finds its place. And with the same stubbornness Hegel, warns that the second part of his work, which is devoted to morality without ethicity, has as its subject abstract morality, the *false* morality.

Only the third part, the ethicity, in which morals and right identify themselves, is true right and true morals. It is the same structure that we already saw in the *Science of Logic*. Just as the concept (= spirit) is the truth of being and essence, so the ethicity is the truth of right and morality.

> One could ask oneself, why don't we start with the highest, that is to say, with what is concretely true? The answer is: because we want to see what is true in a way of a result, and for that one needs essentially to understand fist the abstract concept. Therefore, what is real, the figure of the concept, is for us what comes next, what will arrive afterwards, although in reality itself it exists beforehand. Our procedure is to demonstrate the abstract forms not as subsistent but as false (Rph 32 Z).

"When we speak about right in contrast with morality and ethicity, we understand by right only the first one, the formal one, the one with the abstract personality" (Rph 30 A).

"The morality, just as the previous element, formal right, both are abstractions; only ethicity is the truth of both of them." (Rph 33 Z).

As an interpretative question it is convenient to notice that in 1806, while writing the *Phenomenology*, Hegel did not take yet the terminological decision that we have seen he has made in the Philosophy of Right. In the *Phenomenology* morality is still superior to ethicity. The latter was not a technical term yet. It starts to be so from 1817 onwards, in the Hegelian review of a work of Jacobi (cf. NH 451). That is to say, four years before the publication of his *Philosophy of Right*.

Any serious researcher, as the excellent Lauer, must agree with this: "*Sittlichkeit* simply does not have the same meaning in the two contexts" (1983, 6 n.4). "*Sittlichkeit* on a higher level, treated in the *Philosophie des Rechts* but not in the *Phenomenology*" (1982, 180).

In addition, Hegel himself not only announces the change of meaning, but he addresses it thematically, with which our interpretative question gives way to the question of content:

> Consequently, what we contemplated before with the Greeks as a form of ethicity can no longer have a place in the Christian world. Because that ethicity is custom without reflection; on the contrary, the Christian principle is the interiority that subsists by itself, the soil where the truth grows. Against the principle of the subject freedom an ethicity without reflection cannot be carried out anymore (WG 746).

About the ethicity of the Greeks Hegel says insistently: "It is the ingenuous ethicity, *not morality yet*; the individual will of the subject is located in the not-mediated custom of rightness and laws" (VG 249) (The italics are mine).

"But besieged by immediacy, the freedom of the subject is only custom, without the infinite reflection in itself, without the subjective interiority of *conscience*" (EPW 557).

The following text is particularly explicit:

"For the beautiful ethicity is not true ethicity, it has not been born out of the struggle of the subject's freedom" (VG 250).

It is perfectly clear that the spontaneous ethicity, the one consisting in custom, is not true ethicity for Hegel. True ethicity, which is superior to morality, is the one that the *Philosophy of Right* studies.

"Consequently the form of ethicity is completely modified. The beautiful ethicity is no longer present. What is now ethical, might be also custom or habit, as long as it comes from interiority; but precisely what has plain right is the interior, the subject" (WG 746s).

As can be see, Hegel expressly warns that the meaning of ethicity changes completely. To think, as some Marxist interpreters have wanted, that Hegel's ethicity means custom and habit, would not be an analysis of the texts but a willful introjection of the interpreter's thinking, a recourse of those who, on the one hand, deny the imperative and true obligation as such, and on the other, want to have Hegel on their side.

"But ethicity is duty" (VG 115).

And by contrast, "in the ancient form, the ethicity is custom, habit" (WG 115).

In the entire Hegelian philosophy, in his philosophy of right and of the State, in his philosophy of history, as in his philosophy or art, as in his philosophy of religion, the imperative character of ethicity is of primal importance, because that imperative is God, the only true God. Without that, Hegel could not sustain that the State is the Kingdom of God, or that God is the one who has been conducting the human history and continues to do so.

> In the classic arts, the oracles have essentially their place, because in them the human individuality has not climbed up yet the hill of interiority in which the subject takes out from himself the decision of action. What we call conscience in our sense of the word does not find a place in the classic arts. (Asth I 489)

Let us have in mind that our translation distinguishes between conscience (*Bewusstsein*) and consciousness (*Gewissen*). This last one is the faculty with which we perceive the moral and ethical imperative.

> No matter how beautiful, attractive and interesting the Greek ethicity looks, it is not, however, the culminating point of the self-consciousness of the spirit; it lacks the infinite form, precisely the reflection of thinking in itself, to free oneself from the natural element, of the sensitive that is inherent to the character of the beautiful Greek gods, to free oneself from the immediacy in which this ethicity is located; it lacks self-understanding of thought, the infiniteness of self-consciousness; it lacks that everything which must have validity as right and ethicity finds a confirmation in myself, in the testimony of my spirit (WG 639).

"Men can behave themselves instinctively according to their customs and traditions" (PR II, II 198).

> The Greeks did not have conscience. What is right and duty was the law of the State , with regard to which no one reflected whether it was in fact right and duty. But one is not a free man if one does not *see* that what the State demands is good. And if he does see, then the individual becomes separated from what is custom; the interiority and the formal separation from the existent harmony begins (VG 263).

In contrast with the instinctive ethicity that exists as a general rule in the masses, the true ethicity that matters to Hegel, essentially includes self-conscience and self-responsibility; for this true ethicity the customs and positive laws are not obligatory just because of the fact that they are customs and laws in force; the free spirit demands to understand the purpose of those laws and judge them; one only decides to observe them when one's conscience of good and evil testifies that they are obligatory. This is why the ethicity, far from being a custom, is a 'Doctrine of Duties', as the *Philosophy of Right* calls it when it speaks about the ethicity (142-157). "An immanent and consequent doctrine of duties cannot be anything else than the development of *relations* that are by the idea itself of freedom necessary and hence *real* within the State in its whole extension" (Rph 148A).

Hegel is terribly explicit when he affirms that true ethicity is duty. There can be no doubt about that:

> The difference between the actual exposition and the form of a Doctrine of Duties is only that, in what follows, the ethical determinations are deduced as relations that are necessary; *apodosis* is not yet added to each one of them; therefore, this determination is a duty for man (Rph 148 A).

So, what is the difference between ethicity and morality?

We sustain that a morality, whose criterion and goal is the subject's self-perfection, is rigorously immoral; we sustain that the unique duties are the duties towards the neighbor; we sustain that only the authentic imperative can only emerge from intersubjectivity, and that inventing other kind of obligations that distract us from this one is an immorality, not morality.

The attack is not directed against Kant only, but against all the traditional moralists that have placed the Kingdom of God in another world, with which not only my neighbors and the State become mere means, but also morals itself become subjectivism, since the only thing that matters is the soul that will go to the other world and whose fate will depend entirely on its self-perfection: "My intention of the good of my action, and my conviction that is good, *makes it good*" (Rph 140 A). One has abandoned the *objective* morality that commands to build here a world of justice, freedom and human dignity; and since morality does not become objectively real in the world but is only a soul's garment that only God himself knows, it depends exclusively on my intention; my actions are good insofar my intentions are sincerely good, regardless of any objectivity. "Sheer chitchat: it is good because I am convinced" (Rph Notiz zu 140).

Ethicity means *objective* morality. Therefore, as we said, the only true morality.

We have just seen that ethicity is the doctrine of duties. These are grouped in three areas: family, society and State —a State that is essentially bonded with universal history —. That is in itself objective. The *rights* of our neighbor in each of these spheres constitute duties for me; this is why ethicity is the identification of morality with right. But not by means of an 'also', not by juxtaposition, "not by a mixture of both principles, but by suppressing them and constituting the absolute ethical identity" (JS 509), because as we will see, right without ethics is not true right.

According to Hegel, Kant's contribution not only is irrevocable, but constitutes the true point of reference in the history of moral thought.

Before Kant, no theoretician knew what morals were. The difference between categorical imperative and conditioned imperatives is a paradigmatic achievement: who acts by self-interest, regardless if that means a beatific vision or eternal torments, is not acting morally but only in terms of losses and gains. That conditioned imperative is not really imperative, but only technical information about efficient means. 'If you want this, do that'. On the contrary, 'thou shall not kill' is a categorical imperative and its validity does not depend on one's interest in that commandment; it does not care about whether it is convenient or not for you; it does not care about prizes or punishments. That, and only that, it is moral.

Kant understood that morality begins exactly there where the pursuit of self-interest ends. That is certainly impressive. What he missed to realize was that only the pursuit of the good of the others brings to an end the pursuit of self-interest. I cease to be the most important thing in the world only when my neighbor is as important as myself. That is what Christ's teaching tells us: "So everything you want man make for you do it to them (Mathew 7, 12). This inclination to self profit cannot be thought to come to an end otherwise. By the way, the imperative obtains in that very moment its *content* and becomes perfectly determined. The Kantian formalism, its incapacity of generating real and concrete obligations, comes from not realizing that the imperative comes from the *other*, not from the same subject; e.g. it comes from intersubjectivity. When it comes it comes with content. Building the duties from the inside is a whole pseudo-problem that Kant invents in order to have something to do; he wanted to start from the *abstract* imperative, but that is not a true imperative; it is only the abstraction of the imperative. We will come back to the question of the autonomy of reason in short.

The best commentators already made clear that the acceptance of the superb Kantian contribution by Hegel is total and without reservations. Joachim Ritter says: "In Hegel the adoption of Kantian morality in the *Philosophy of Right* has a fundamental importance" (in Riedel ed. II 1975, 218). Likewise, W.T. Stace says: "Thus the Hegelian ethic contains all that is true, good, and noble in Kant's ethical system without its defects." (1955, 379)

Hegel himself expresses it unmistakably: "The merit and the elevated point of view of the Kantian practical philosophy consist in having highlighted the significance of duty" (Rph 133Z).

"This discovery of the absoluteness of reason in itself, which has produced the turn of the philosophy in our times, this absolute point of beginning, must be recognized and it is irrefutable, in spite of the fact that one must call insufficient the Kantian philosophy" (Ästh I 109s).

The absoluteness of conscience is an indubitable fact: the sharpness of the imperative is only perceived in the moment when I convince myself that the content is obligatory.

Despite that the Kantian contribution is, like Hegel says, an absolute starting point, the insufficiency of the Kantian system must still be affirmed: "incomplete morality [...] is immorality" (PG 440).

> No matter how essential it is to highlight the unconditioned self-determination of will as the root of duty —in fact our knowledge of the human will only has its firm fundament and beginning point due to the *Kantian* philosophy through the idea of its infinite autonomy —this acquisition still turns out to be a *formalism* if one holds it only from the very moralist point of view, without arriving to the concept of ethicity; the moral science becomes mere rhetoric about *the duty by the duty itself*. On the basis of this principle a Doctrine of the Duties is impossible; [...] from the notion of duty as absence of contradiction, as formal coherence with itself, which is nothing but the attachment of the *abstract indetermination*, it is impossible to arrive to the determination of concrete duties (Rph135 A).

"Although we stressed above the point of view of the *Kantian* philosophy —which is excellent insofar it affirms the conformity of duty with reason—, one must still denounce its mistake, namely, that this point of view still lacks a whole structurization. For the sentence 'May you live your life as if the maxim of your actions were to become universal law' would be very good if we had already determined principles about what had to be done. Indeed, when we demand from a principle to be the content of a universal legislation, it must have beforehand content; and if the content was there, the application would be very easy. But here the principle itself is not present yet; and the criterion about non-existence of contradiction produces nothing; where there is only nothingness we do not find contradiction" (Cfr. Rph 135 Z).

Kant has to employ certain artificial resources in order to have content —at the end of the day these resources leave him in a bad situation. As Hegel sharply criticizes, to examine if something can be erected as universal maxim, presupposes that this 'something' has content, and therefore the existence of a content is not originated in such inquisition.

In the same way to know if something can be held without contradiction, first we need that *something* to have a content. So the existence of the content is not originated in that research. The notorious artificiality of these resources allows one to see clearly what we were saying: Kant seeks that lack of content because he wants to. It is false that Kant has in front of him the imperative and its lack of content. 'An imperative on the quest of content' it is something artificially fabricated by *abstracting* the real and concrete imperatives to the pure notion of the imperative, which, in fact, results to be the *abstraction of the imperative*, and this has not the exigency and is not anymore an imperative.

Now, this entire Kantian detour has the objective of saving the autonomy of reason. But we have already seen (V,1 fine) that the autonomy consists in the decision of the subject itself , and not in the natural impulses that where already there. The ethical demand that addresses the neighbor is what makes this autonomy exist; the imperative that stems from intersubjectivity is what frees man from the natural impulses. There is no autonomy of the will without the moral demand that the others address to me. Autonomy as the starting point with no concrete duty is an abstraction, the abstraction of autonomy, and therefore the entire problem of conciliating the autonomy with the existence of particular obligations is a pseudo-problem.

Hegel refutes, therefore, the Kantian system from the inside, that means to say, by adopting the complete univocity of the absolute imperative without mixture of conveniences and self-interest, but showing that the formalism contradicts the origin itself and the existence of this imperative is the same as immorality. What Hegel does is to "establish the true principles of morality or rather of ethicity against false morality" (VG 171). Referring to Kant and Fichte he affirms: "Everything stays in this gossip about morality. But *what* things are moral and how will the spirit realize itself systematically is something no one thinks of" (GP III 369). It is in the rights of the neighbor where one establishes what things are moral and what things are not. A system that pretends to establish duties, distracting us from our duty towards the neighbor, is an immoral system.

It is obvious that Hegel does not refute only Kant and Fichte. It is a widespread weakness both in moral and juridical treatises to distinguish between morals and right, by saying that the former only pertains to the interior and the intentions, while the former does not care about such things.

Just as we have seen in the previous chapters, that physics, chemists and biologists live, by common agreement, in the fiction that their disciplines are empirical, so jurists live by common agreement believing that right does not care about intentions or the interior but only about the exterior.

This is a negative universal thesis. Consequently, one needs only one particular case in order to demonstrate that it is false. Now, in the branch of law called criminal law, the interiority of the accused is a decisive factor for the code itself, the judge, the accusing instance and the lawyer. If the murderer is insane, there is no crime. But insanity or sanity pertains to interiority.

All the criminal codes of the world distinguish today between willful harm and negligent harm, that is to say, between a prejudice caused with the intention to harm and a prejudice caused by imprudence. The felony is completely different, and the penalty, consequently, is also different. The intention of the people is something fundamental for the law.

The criminal code of Mexico City, in its article 225, fraction VI, imposes suspension, dismissal or a fine to the civil servant that dictates or omits a resolution that violates any categorical precept of the law, "whenever one acts by immoral motives and not by simple mistake of opinion".

This applies not only to criminal law: this is also valid for civil right and commercial law, since one looks there for intentions as well. For example, a testament is invalid if the testator did not have the intention of making a will. Also, civil right codes invalidate a contract when any of the parties suffered a 'mistake over the determinant motive of will', as jurists frequently claim. That is what the Napoleonic Code called a mistake over 'the cause of the contract'. And the invalidating mistake could have also been related with the nature of the contract.

When a judge has to decide which member of a broken marriage will take care of the children's education, what matters the most to him is the interior personality. The father's lawyer intends to show the psychic instability of the mother or her internal viciousness. The exterior facts function only as indicators, symptoms or proving means, and the truly decisive for the judge is the interior. It is completely false that in these cases the right is mostly concerned with the exterior and accidentally the interior. The thing is actually the other way around. What matters is the internal personality of husband or the spouse. And the same is to be said about the testament and the contract: the decisive factor is the intention or the interior experience of mistake or ignorance. The exterior data are equivocal: in the case of a person certain external data can be an indicator of ignorance, and in the case of a different person the same external data can be irrelevant. In the examples I have mentioned, right does not care about the exterior in itself, but only as a means to hit the mark with the interior.

Let us think now in the felony called fraud: only the interior has importance. As far as the victim goes, the decisive thing is to find out if he was effectively deceived by the fraudulent maneuver and not by his own imprudence. As for the felon goes, the decisive thing to examine is the intention of making the other fall into the trap. No exterior data has univocal significance, since the same fact can deceive certain individuals under certain circumstances, but that may not always be the case. This demonstrates that it is not the exterior fact what orientates the right in those cases, but only the interior, because sometimes it can be accompanied with certain exterior facts and sometimes with others very diverse. The criminal lawyer Mariano Jiménez Huerta recognizes with honesty: "The deceits cannot be measured objectively, because, projecting themselves over the intelligence of victim, they will necessarily influence in its efficiency the subjective psychological peculiarities of the deceived person" (1 968ss IV 140).

The examples we have seen solve the question without any doubt. The ideology that prevails nowadays among the jurists —which 'justifies' judicial, legislative or governmental decisions that are flagrantly unjust by saying that right has to relay on the exterior and the moral character of the facts does not belong to legal right— is completely outrageous. Insofar one is a lawyer or a judge, one feels exempted from the universal moral imperative, but the pretext they come up with destroys itself: how do they justify the decision they have taken, namely, that the exterior is in some cases relevant and in others not? The few examples we have mentioned demonstrate that it is utterly false that the interior is irrelevant to right. That being proved, how can they justify to do without morals whenever they want to?

Hegel refers to this in his *Philosophy of Right* when he says that the positive right, that is to say, the abstract right, is not true right, and that true right is only the ethicity in which the morality and right identify themselves. Already in his youth Hegel condemns this: "It may also happen that a conduct of right is not bound to the mentality of right, and even that there is an immoral mentality instead" (NH 253). Here we read a criticism against the hypocritical cynicism that has found today a 'scientific' rationalization in the juridical positivism of those who wield the law to commit, knowingly, atrocious injustices.

"An action like the affirmation of my property can be perfectly juridical and yet there can be intention of evil, regardless of right and having the purpose of harming the other" (NH 228).

"When in politics, the only thing that prevails is law as such, the dominant thing is violence and the arbitrariness of the individual" (PR II, II 82s).

"The highest thing, if conceived finitely, is the worst" (PR II, II 238) [...] "an animal kingdom of men" (ibid).

"Thus in right the particularity is not yet that of the concept, but only that of the natural want" (Rph 151 Z).

In order to do without the obligations of the moral imperative when it comes to the law, jurists have relied on a huge epistemological mistake: the mistake of believing that the application of a law is a mechanical process that may be performed without judgments of value, without appreciations that are empirically unjustifiable. As a matter of fact, it is very unusual to find jurists that take the task of reflecting about the epistemological status of their own acts.

As we have seen, physicists and the biologists fall in that ingenuity as well, and they figure that if 'the whole world' thinks in a certain way, that is enough to say that what we have in our hands is an empirical judgment. Only that in the case of the jurists that lack of reflection has ominous consequences, for it has lead them to believe that the action of subsuming certain empirical fact under a law does not imply valuation and *decision*, because the 'whole world' would subsume the fact in question under that law. But the valuations do not stop being valuations because they are common or very frequent. To be sure, not even the sentences that allegedly describe a fact, that is to say, the so called 'fundamental propositions', are justifiable only by empirical data, since they *decide* precisely which is the content of the supposed empirical data. The empirical data does not consist in words; the descriptive sentence, on the contrary, consists in words. To judge if the sentence reflects 'sufficiently well' the sensorial data is always an appreciation, an evaluative estimation, not a mechanical process at all. The vocabulary with which one supposedly describes the fact in question decides beforehand under what law it will be subsumed. I am not reproaching this way to proceed; on the contrary, I am saying that it is the only one possible and that is the way one always follows. But the point is this: if the judge, the ruler or the jurist in general carries out judgments of value even with the actions which are apparently more mechanical, if they are always carrying out moral judgments, how could one do without the moral imperative by saying right does not consider it? With what logic or justification can someone base himself on right in order to commit acts and deeds which his conscience unmistakably calls injustices and immoralities?

It is not the case that the unjust jurist does without morals. The important thing here is that he embraces immorality. And in that moment, his decisions are "the oppression of right" (GP I 142), "his right is a right that lacks right completely" (Ästh I 305).

The reader may have recognized among these Hegelian expressions the unmistakable thesis of iusnaturalism. We shall demonstrate that thesis is true.

The juridical positivists have not noticed yet that the word duty lacks entirely of meaning if one does without the moral absolute imperative.

In the juridical jargon, the 'objective' right is synonym of law. On the contrary, the 'subjective' right is the faculty or power of making something. One should notice that the possibility of giving any meaning to the subjective right depends on that objective right has meaning. To say that someone possesses the right of something is tantamount to say that the others are under certain law or duty in relation with him. Thus, if the word duty or law does not succeed to get meaning, the word right lacks of content in its objective acceptation because in it the right it is a synonym of duty or law, and in its subjective acception because the existence of a faculty in a subject *means* only the existence of a duty in the others.

It has been a superficiality to forget (cf. III,9) that historically speaking, all the people supposed that it was a divine imperative to obey the laws, whether they were written or not; it was considered that the laws were originated by the divine will, regardless of the particular mediations they may have had. This is the only reason why the idea of duty or obligation could be once related to the effective precepts in a human conglomerate. Besides, it was assumed that every authority emanates from divinity.

Without that original meaning it would be absolutely impossible to apply the word 'duty' to merely cognitive information about the real results of certain kinds of actions. For instance: 'whoever makes such thing would be whipped'. Kelsen seems to ignore the difference between categorical and hypothetical imperatives. As a matter of fact, the latter is not an imperative that is to say, it is not a duty; it is only some sort of knowledge. If I do not want to be whipped, I refrain from carrying out such kind of acts, because there is another agent who will apply me that punishment.

The same happens when I refrain from getting my hand close to the flames because I know that in such a close distance I will suffer a burning or some kind of pain. But I do not have a duty that prohibits me to draw my hand closer to the flames. From that kind of knowledge no duty can be derived; I can only infer from it how *convenient* it is for me not to do certain things. No one could ever come up with the idea of calling duties the threats of the strongest. The persons who are threatened by someone with a machine gun do not call duties the indications they are given by him; obeying them is simple reasoning of advantages or disadvantages.

This is like a joke, but many positivists believe that because 'the whole world' calls the statements of a positive legislation law and right, they think that law and right are empirical data. But one needs a bit of reflection to understand that the *obligatory* character of such statements is by no means empirical data; consequently, the fact that such sentences are law and right is not empirically perceptible, and therefore the existence of a law is not empirically verifiable. Kelsen, fortunately enough, does not fall in such ingenuity, so we can leave that problem aside.

We will prove this afterwards with a quote. It is important to have in mind that Kelsen says innumerable times that right is not sensible data. Kelsen affirms also that the anarchist who says right is nothing but the power of strongest is irrefutable, and that one cannot justify the hypothetical decision of the jurist and the theorist —Kelsen included— who regard as right the prevailing order of a nation. And it is obvious that Kelsen omitted asking himself what, then, is the origin of the idea itself of right. The precepts or rather the sentences empirically perceptible do not raise in the human mind the idea of right. On the contrary, this is a contribution that the jurist introduces in order to look at them as right. That being said, where does that idea come from? How did it originate itself? Evidently, the only possible origin is the moral absolute imperative, the one that positivism eagerly denies.

One last thing must be well considered. The constitutive thesis, the only essential one of juridical positivism consists in holding that there are no more duties than those consigned in the juridical positive order; to affirm that there are others is the constitutive thesis of the iusnaturalism, so that the positive laws are only true laws if they coincide with the moral law or are logical derivations from it.

Now, if the duty that the categorical imperative imposes was illusory, it would be inexplicable that Kelsen and his comrades call it illusory, since such qualification can only be given by contrast with a true duty, and the idea of a true duty does not arise from the empirical data or from the juridical positive order. The only reason why positivists can still distinguish true and illusory duties is because they have awareness of a duty that is not illusory. This demonstration is definitive.

Only if one assumes that it is *morally* compelling to obey the ruler and the rules of that country can these be called laws, duties and right. And if people call them so, independently from the decision of the jurist, it is because they originally made such supposition, so that even by inertia one preserves such denomination in the entire world.

The jurist that denies the moral imperative does not have the right of using it, since without the imperative words like law, duty and right lack all kind of meaning. Kelsen mentions Kant several times, but, apparently, in the *Metaphysics of Morals* he overlooked this: "From the *moral imperative* which imposes us a duty, the concept of right can be developed afterwards, which is the faculty of obligating the others (op. cit. AB 48).

The concept itself of right only obtains content to the extent and degree in which it coincides with the moral imperative. Otherwise, as Hegel says, "instead of right it is oppression of right" (GP I 142), "it is entirely positive and with no validity or force" (Ästh I 305).

Saint Thomas expressed this with the same clarity, although with a vocabulary that we will criticize later with Hegel: "Every law put by men is a law insofar it is derived from the natural law. If it differs in something with the natural law, it is no longer a natural law but a corruption of it" (I, II 95 2).

Before we move on forward, we shall document with words of Kelsen the non-empiricity of right: "The fundamental mistake consists in believing that the social order is a natural right, perceptible in the world of causal reality" (1965, 350); "in the kingdom of nature there are no 'robberies', 'juridical businesses, 'punishment, or 'execution'" (1965, 62).

> The *Pure Theory of Law* does not try to dissimulate that the specific normative signification of the so called juridical facts is the result of a possible interpretation, not a necessary one, which is linked to a fundamental hypothesis. Furthermore, it realizes that it is not possible to prove the existence of Right as if it were a natural phenomenon, or to refute with decisive arguments the conceptions of the theorists of anarchism, for whom the right that the jurists talk about consists in pure and simple force (1968, 99).

> [...] one cannot force an anarchist to recognize that the facts that from the point of view of the state law are interpreted as right constitute juridical acts and not acts of force. One can only demonstrate the possibility of understanding the set of constitutive facts and norms of international law within a unitary system in which all juridical and state orders are contained. The material that is interpreted as law does not constitute a priori 'right', but it rather requires that quality by means of the juridical hypothesis which is useful to interpret it. (1965, 168)

As we have said, Kelsen did not consider that, in order to be true and have some meaning at all, the hypothesis itself needs first to have *a meaning* and be understandable. Since meaning does not have its origin in empirical data, it can only have its origin in the moral imperative. As Hegel says, right is something sacred by definition; no material or physical data gives raise to the idea of right. This idea is far beyond everything which is empirically verifiable; it is impossible to conceive earnestly the meaning of the word 'duty' without some sort of sacred. Kelsen himself emphatically insists in the fact that it is not sufficient that certain acts or statements emanate from a group or a person called legislator or gubernator to be called law, neither it is enough that they do not contradict constitution, for such a group, or a person, or the constitution itself would be first to be law, something which cannot be empirically corroborated.

Although the superficiality of those who believe that right is an empirical data, has been refuted, I would like to draw the attention of the reader to a point in which this superficiality has proved to be particularly stubborn and even childish: the right or property. On the one hand, the defenders of property wish it to be an empirically verifiable data in order to affirm that it exists before or independently from the State. On the other hand, it is absolutely essential for Marxists that property —or at least possession— is a physical and material fact, because otherwise the thesis that the relations of production —which are relations of property that Marxists call 'infrastructure'— determine the ideas of men would not be materialist.

Jurists have written thousands on volumes about how constitutive property is. Savingy and von Ihering have devoted a lot of their efforts to this issue. However, we are not going to quote any of them in order to solve substantively the question.

The fact that I am the owner of this typewriter machine is evidently not a physical nor a material fact; it does not consist on whether I can hold it in my hands or not, nor the fact that it is within my house, nor that it is in the surroundings of my body since I can lend it for many years to a friend, or leave it in a workshop in order to be repaired, and despite of that, I would still be its owner. Property is ideas. Property consists in that I have the right to determine what should be done with the object; in other words, the others have the obligation of not hindering me from doing with the object whatever I please.

But not even possession is a physical data. As a landowner of a field, I can be the possessor and yet not the owner; perhaps I live many miles away from it, while the other landowner has more physical contact with the field than myself. The same goes for the workers that plow my land. Despite of that, I am still the owner of that field. Any person who walks by it has the same *physical* possibility as myself to work in that piece of land, or even of destroying it by placing chemical substances in the ground. This makes it clear that possession is also ideas. It consists in the right of using the object in the ways its owner commands.

Despite the intention and the wish of Marxists, the constitutive thesis of Marxism does not have the slightest trace of materialism —and by the way, it is a false thesis, but we will not deal with that here—. The relations of property are the only decisive element in the mode of production: again, we are dealing with ideas, not with facts. To sustain that ideas are determined by ideas is a triviality that does not deserve our attention.

Let us address now another difference that treatise writers believe to have found between morals and right: the coercion of the latter. The principal difference, namely, that right pertains to the exterior and morals to the interior, has been refuted. They intend now to tell us that right is essentially coercive, that means to say, it recourses to coercion in order that people respect its precepts, while morals does not have that characteristic. Therefore, they affirm, there is a true distinction between morals and right.

But this argument is completely thwarted. On the one hand, there are many norms that these very treatise writers would call merely moral and yet they are accompanied by a social sanction whenever they are not respected. On the other hand, there are not only many juridical precepts that *necessarily* remain without a sanction: the coerciveness of right is sometimes reduced to the *moral* permissibility of the use of coercive force.

Anyone who reflects a little knows that the liar is sanctioned with the collective rejection of everything he says and even with the contempt of his very person; the cruel person is sanctioned with isolation and with displays of horror that end up being a boycott against him; the deceitful person is excluded from taking part in businesses. The list is practically endless.

Those people —who, in view of the immorality that prevails nowadays, do not regard highly the effectiveness of these sanctions— take no notice of two important things. First, they are imagining that the legal and juridical sanctions have an extreme effectiveness, disdaining thereby, the opinion of many sociologists who think that *most* of the crimes remain unpunished. Second, they do not realize that saying 'moral precepts do not have any corresponding sanction' is a universal negative proposition, which only needs a *particular* case that proves the contrary to be refuted. With the examples we have mentioned we do not need to enlist all the sanctions that go along with all moral precepts.

But whoever considers things carefully will realize that, in fact, the 'social order' prevails not only due to the sanctions contained in the penal code, but to a much greater extent due to the sanctions that are not contained in it.

And this goes not only for underdeveloped nations. One would only need to look at works such as *The Disputing Process* (*Law in Ten Societies*) (1978) written by Laura Nader and Harry Todd Jr., and the investigation of S. F. Moore called *Law as Process* (1978) to realize this. Not to mention the works of Sugarman, Fitzpatrick, Pospisil, etcetera.

In this context it would be useful to take a look of this issue from an anthropological point of view. Let us quote that Ino Rossi makes of Radcliffe-Brown findings:

> Radcliffe-Brown points out that in many primitive societies there are no courts or judges, nor any formally organized, central political authority, —but nevertheless people possess a sense of wrong-doing and the notion of public and private delicts. Whenever there is no political authority to apply organized sanctions (law), there are private groups and associations that apply organized sanctions. Moreover, there exist a series of unorganized mechanisms like ostracism, house-burning, blood feud, accusation of witchcraft and sorcery, and ritual sanctions which although not enforced by official authorities, are effective means of social control. (Rossi *et al.*, eds., 1977, 348)

As for the developed countries, Nader and Todd refer to the investigations of Macauley and Sutherland, in which they describe: "avoiding the law as a means to form and maintain good business relationships; businessmen prefer not to use contracts, and even prefer not recurring to the law in the cases of criminal business activity" (1978, 17).

Let us only think about this: to lose one's prestige in the business world, the avoidance of any kind of deal with the transgressor, the rejection of providing him with credit and resources, are for a businessman sanctions which are infinitely more harmful than those prescribed by the law. The bills of exchange and promissory notes were in the 16th and 17th centuries a matter of honor and prestige for the traders. And even nowadays, the capitalist national systems would collapse without the trust in verbal agreements. It is obvious that even in our times the relations between capitalists cannot be based on the legal or juridical system: a juridical process is long and one can appeal its results, and at the end of that course of action my business would be surely broken or I would have certainly lost the possibility of doing the business move I had in mind. The only trust I can have in those scenarios it that which the verbal agreements can provide me: my own word.

Besides, in the little towns and in the neighborhoods strongly bound that we find in the big cities, the sanctions that are not legal can make life impossible for everyone: the mailman, the plumber, the gardener, the carpenter can all cooperate to impose me a sanction by not providing me some elemental services. It is impossible that the positives argue that the legislator 'implicitly' approves and assumes the social sanctions, for many of them contradict the law itself.

The facts refute the thesis according to which the moral law does not have sanctions. We will later relate this fact with another idea. But now, let us concentrate on the second part: the one that affirms that law is essentially coercive.

To begin with, we have international law: a law which that does not have a coercive character like many treatise writers affirm. Against a powerful nation that transgresses the international juridical norms there is no sanction, or at least, not a sanction that is different from those which are characteristic of moral norms. A superpower can be more powerful than the United Nations and the Permanent Court of International Justice.

And let us not forget this crucial point: there can be a criminal in each country stronger or more astute than the several governmental organizations and who remains hence unpunished. The robberies and aggressions that happen all day long are an evident proof of this; the most elemental realism tells us that many of these crimes remain with impunity. Therefore, in what does the coerciveness of law consist? Would one say that it consists in the fact that the law affirms that those crimes *deserve* a punishment? *That is something which morality tells us as well,* and the result that follows thereof that there is no difference between moral and right.

The right of self-defense, which is explicitly recognized by the majority of legislations, cannot mean the *real* available force. The victim has this right in spite of the fact that his aggressor has stripped him of the possibility of defending himself. In this case we have a right without having effective coerciveness at all. The theorist that denies the victim this right by the circumstance itself that he is being attacked in a highly effective way, what he is doing in fact is admitting that force, far from being inherent to right, suppresses it. On the contrary, if the theorist affirms that in this case the right of self-defense subsists, he recognizes thereby that right is not essentially coercive, and consequently, it does not distinguishes itself from morals. As a matter of fact, the right of legitimate self-defense, which is acknowledged by every treatise writer, consists exclusively in the power of the victim to use violence in a *morally licit* way.

The thesis that we have been criticizing could not reply that the aforementioned authorization depends on that the positive law authorizes the victim to use violence, for that very positive law would not be coercive. Again, morals and right cannot be considered as separate things.

Aside from legitimate defense, there are at least two cases in which the right *necessarily* lacks coactivity.

The crime of not denouncing a criminal has to remain unpunished, because otherwise we would be in a *processus in indefinitum.* In the first link of the chain, the penal code speaks of penalties, but, what about the crime of not denouncing someone who commits the crime of not denouncing? This crime has to remain without a sanction, because otherwise one would have to sanction he who does not denounce the people who has not denounced, etcetera.

Necessarily, the crime of not denouncing someone who does not give away a delinquent must remain unpunished. One should notice that in the real existing codes there are many laws which forget to determine the penalty to the crime they prohibit, but the defenders of the thesis we have been criticizing would respond grandiloquently that these laws are not right and period. The argument we have been discussing, however, does have nothing to do with that. It is not about crimes which have remained unpunished by accident, but of a crime which, due to the very logic of things, will necessarily remain without a sanction. Consequently, it is false that right is essentially coercive.

As the thesis we have been refuting is a universal affirmative statement, one only needs to demonstrate that it is completely false. There is another example we can bring up in order to reinforce our point: the juridical order cannot determine a sanction in case in which the head of state commits a crime and resists the sanctions prescribed by law. The juridical order would have to create another instance which is authorized to employ more force than the one it originally granted to that head of state. That scenario, however, could repeat itself again in an endless process.

The result of the above is that the juridical order necessarily contains some precepts without a sanction. If it grandiloquently responds that those precepts are not right but morals, the very same thesis they are defending falls apart, because it would turn out that the juridical order *necessarily* contains some moral precepts and hence is false that right is different than morals.

This convergence and identification between right and morals has been our very subject: the Hegelian ethicity which is the truth of both right and morals: "Neither the juridical nor morals can exist by their own, they both need to have the ethical as carrier and fundament [...] right only exists as a branch of a whole, as the plant that grows in a robust and independent tree" (Rph 141 Z).

"Imagination figures sometimes that the State remains united due to the force, but what sustains it, is the fundamental sense of order that everyone has" (Rph 268 Z).

Positivists make a mistake of imagination when they do not realize that the actuality of the State is ethics and that there are innumerable ethical duties that one forcefully and logically needs in order that the legislator or a governor can give a command. Let us think of the duty of not lying, of the duty of taking the man that addresses me as a person and not as a bunch of atoms, of the duty of giving sense —to the extent I can — to the content by itself chaotic of sounds that the person in front of me emits, of the duty of assuming the intersubjective obligations that every 'verbal action' implies.

Let us stop for a moment in this last point. Both the speaker and the hearer must assume specific duties, which depend on the specific verbal action we are dealing with: promises, advises, regards, threats, etcetera. It is impossible that us humans understand each other if we do not hit the mark with each kind of verbal action that other speakers are trying to carry out, if we do not adopt the corresponding role; for instance, if we do not realize that what the other person is saying is a yoke, we are in fact not understanding anything. But for each kind of verbal action some specific duties follows, without the acceptance of which the verbal action could not have been possible.

If I made a promise, the duties that follow from that are obvious. But the same goes if I gave a piece of advice: I adopt therewith the obligation of not being surprised if the other person behaved in accordance with what I said. If I made a plea, I should not be surprised if one proceeds in the way I suggested. Besides, in that case, I also acquire the obligation of being thankful. If I made a statement, I am obligated to response for the logical consequences of what I said, and I am obliged to concur when somebody says the same thing I did. Every verbal action constitutes an ethical fact that has consequences: both the agent and the receiver who accept to participate in them assume their consequences. For instance, the person who was given an advice has the obligation of not being surprised of the negative outcome if he acted contrarily to what his friend said.

And not only are the duties related to language the ones which constitute the complex that we call society and state. Let us think in the duty of not aggressing each other; in the duty of not making fires; in the duty of producing the goods and services of what I am responsible; in the duty of behaving in a way that I do not hurt anybody; in the duty of prudence and moderation of movements so that my neighbor can live calmly and in peace, etcetera.

This series of innumerable ethical duties is the *condition of possibility* to every act of authority, whether it is legislative, executive or judicial. To think that some governor or legislator *creates out* of nowhere human co-existence is pure lack of reflection. This series of rights and duties is *prior* to everything: it is the condition of the possibility even of *speaking*. Such a hypothetical 'founder' cannot be said to adopt and assume legislatively all these duties by transforming them into positive right, because they are *innumerable*: sociologists and anthropologists even nowadays would not be able to make a complete list. In what serious way can one say that the legislator or the judge make positive a series of duties that do not even know of? Evidently, there is no conscious act on the part of the legislator, the judge, or the politician in relation to these duties. The 'implicit' assumption that the positivist theorist would attribute them would be an act of such theorist and not of them, and therefore it would not be positive right but the imaginative lucubration of the theorist.

International law expert Clive Perry, who has defended the juridical character of international law despite its apparent lack of coerciveness, has rightly remarked:

> No juridical system depends completely, either for its effectiveness, or for the definition and development of the norm it contains, on the application of imposition by means of a superior power. If men decided not to obey the law within these State, there could never be enough policemen that could force them not to behave like that. Exactly the same kind of recourse with which one counts for the preservation of the apparently high degree of order that prevails within the States of the world, are available to preserve the international order as well. Theoretically speaking, they can be employed with no lesser degree of effectiveness in both of these realms (Sorensen ed., 1978, 55s).

As a whole, the displays of social disapproval are a sanction much more efficient than those foreseen by the law and the application of authority. And the same goes for the spontaneous action of the passers-by who try to prevent a homicide or an act of cruelty. The duties attached to the 'verbal actions' have as their sanction the systematic exclusion from the community for those who do not behave well. Hegel says correctly that positive right can only exist as a parasite in another right that is not written and which is called ethicity. No one can deny the character of right to ethicity, because its sanction, taken integrally, is more efficient than that of positive right.

Let us only insist in the non-written precept that says 'you shall not deceive'. As it is a constitutive element of speech, it is evidently the condition of possibility of any positive precept; and evidently, society also punishes whoever does not respect this by boycotting him once his felony is discovered. Plato energetically rejects (Laws VII 793 BC) the opinion of those who think that these 'not written laws' are in fact laws, for they are, as Plato says, the fundament of the written laws and the bond that ties ones with the others together.

It would be completely paradoxical to believe that the precept 'you shall not deceive' does not exist if some legislator does not establish it. Positivism would have no other way out than saying that the entire community was legislator before some individual or group assumed such functions. In that case, there would be no difference between positivism and no-positivism, for there would be no difference between legislating and not legislating, The antipositivism holds that one does not need legislators or authority in order that right exists. Positivism affirms the contrary, but in order to be a legislator one only needs to be considered by the theory as such. *Reality* does not need to be different than the way anti positivism describes it; for the correctness of the positivist thesis one needs only the whim of the positivist theorist. In fact, we deal there only with mental toys and nothing more.

4. State

Rousseau's disastrous mistake was to consider man good by nature. But that man is man by nature has been and still is the *absolute mistake* within vulgar thought, Sociology, Psychology, Politics, Pedagogy, Theology and Philosophy, despite that Plato and Aristotle demonstrated the contrary twenty five centuries ago and in spite of the fact that Hegel deepened that precise demonstration almost two centuries ago.

When Aristotle (Pol I 1253a25) and Hegel (JS 505 *et passim*) stridently proclaimed that by essence the State anteceded man, several thinkers tore their clothes believing than that priority of the State attempted to tell us what man ought to be. It did not even cross their minds that *what man is,* is what it is all about, it is a *confirmation*. In spite of the fact that Hegel takes heed of it since the preface itself: "To understand *what man is*, such is Philosophy's task." (Rph xxi fine)

I already remarked (I, 1) that modern anthropology agrees that Hegel is right: by nature man is not man but animal; the so-called 'human nature', which according to the biased and pedant tyro is 'the same everywhere', is not human but primate.

If only the anthropologists of our century had read Hegel, they would have synthesized their own research in the following phrase: "The fundamental principle is that man as such is not a natural being, he is not an animal." (PR II, I 27)

This is everything those who scream blue murder must know when they read: "Everything man is, he owes it to the State; only there he has his essence." (VG 111)

It does not surprise me that such a good scholar like Mure states that Karl Popper's invectives against Hegel are "blatantly ill-informed" (1965 viii), neither what Ripalda claims about them that "it is hard to read something more idiotic." (FR xiv)

But Karl Marx, an author Ripalda does not take to be idiotic, turns out to be just as incapable as Karl Popper to understand that the State makes the man and not vice versa, even though there is no clearer issue than the fact that by nature man is an animal but not man. It is enough with the next 'critique' (?) from Marx's pen against Hegel to make clear that the former simply never got what the point was: "I am man because I was engendered without society's arrangement; this particular creature that I am can be transformed into a lord or a king just due to social arrangements." (MEW 1 310s) What is claimed here is that man becomes man without society's intervention.

We do not find in Marx or in Popper the slightest suspicion that *zoôn politikôn* (Pol I 1253a3) is the *definition* of man. Both still believe, as every coffee shop philosopher, that in the beginning there was a time in which there was no State or Right, the strictly speaking contradictory time that they called the situation of nature, as if the expression 'natural man' wasn't a *contradictio in terminis*. Notice that I am not claiming that they believe in a myth, even though such time is indeed a myth; I am saying that they have not understood the difference between man and animal. The same happens to Kelsen by the way; since only positive Right is Right, supposedly before legislating and governing was invented there was a 'natural situation' in which man was not man and there was no Right.

The refutation of this superficiality was formulated in the third chapter (III, 7 and III, 9). There, I showed that without self-consciousness man is not man and that the rise of self-consciousness is impossible without the social appeal that demands duties and rights. I beg to confer. Whatever Marx says, no one is man by birth without the intervention of society.

Besides this fundamental error, Marx makes the mistake of believing that civil society can exist without a State, the mistake of not understanding the Hegelian thesis according to which "civil society [...] supposes the State." (Rph no. 182 Z) A thesis Aristotle insisted on relentlessly. The Marxian text I quoted offers the advantage of making apparent that Marx's is a core mistake; that Marx denies the intervention of *society itself* in the constitution of man as man. This fundamental mistake discredits the whole Marxian critique of Hegel's State and Legal philosophy. And it is important to remark that according to Engels (MEW 16 362s) this critique is the origin of Marxism.

Putting Marx aside, if we are to reject the accusations of statism, it is not to calm down its accusers telling them that there has been a misunderstanding and that in the end Hegel claims the same as them. It is not about an issue of terminology, as if after clarifying that Hegel does not understand by State the same that them, everything would be settled. It is not true that anyone can define State as it pleases him. Studying an author neither consists of verifying if he holds what I already knew.

The priority of State seems to them unbearable because *they* identify State with government. They overlook that the identity means Legal positivism because they make the existence of individual rights depend on the ruler. Only who defends human rights that do not depend on any authority can logically reprove statism.

In every single European country there is a real distinction between the head of State and the head of government, which logically implies a real distinction between State and government. A theory conscious of its duties has to define State in a way that government does not get included in the definition.

No doubt the definition proposed by Hegel was crystal clear since the beginning. For example, in his 1830 review, F. J. Stahl summarizes Hegel's ideas as follows: "According to his ideas, with the concept of Right the concept of State is already given; it is enough to scrutinize the former in order to find the latter." (In Riedel ed. I 1975, 223)

In *Philosophy of Right* Hegel defines State since the preface: "the ethical universe" (*das sittliche Universum*) (Rph xxii), *i.e.*, the set of the ethical.

And he said lines above: "The ethical world, the State." (Rph ix) The same definition is found later in the main part of the book: "The State is by itself and for itself the set of the ethical." (Rph no. 258 Z)

We have seen what the ethical is for Hegel: the intersubjective rights and duties free of, and contrasted with, a narcissist morality and a positive Right that would make them depend on legislations. The set of the *true* duties and rights is the State. By the time Hegel gave his legal lectures to the *Gymnasium* students at Nüremberg he already held that: "The State is the association of men under legal relationships." (NH 246) Afterwards, in his *Philosophy of History*, he held exactly the same: "We call State the spiritual individual, namely, the people as structured and transformed in an organic whole." (VG 114)

The significance of both the government and the monarch are so diminished that the monarchist K. E. Schubarth attacked Hegel's doctrine in 1839 with the following words:

> The prince is not the substance of the State, which is really constituted by the set of the different particular organic spheres such as family and civil society according to its structuring in diverse social estates, corporations and chambers. All this makes the prince's importance completely accidental. (In Riedel ed. I 1975, 254)

Both those that accuse Hegel of statism and those who with huge ease label him the ideologist of the Prussian monarchy should know that Schubarth's work is entitled: *On the Irreconcilability of the Hegelian Doctrine of State with the Supreme Principle of Life and Development of the Prussian State.*

Schubarth's fears were not groundless. Hegel expressly says:

> For being a monarch it is just required a man who says 'yes' and who dots the i's and crosses the t's; since the tip should be such that the particularities of character are of no importance. Beyond this last decision, any other property the monarch possesses ought to be reduced to peculiarities anything could properly rely on. It is true that there might be stages in the development in which those peculiarities might stand out, but in that case we would be dealing with a State not fully developed yet, which is not well built. Within a well-ordered monarchy *only law is concerned with the objective;* what the monarch adds is just the subjective 'I want' (Rph no. 280 Z) (my emphasis).

It should not surprise us this kind of marginality of the monarch (and of the government itself, as I will say immediately) if we recall that since the preface we were warned that the State is the set of the ethical: "the rich structuring of the ethical in itself, which is the State." (Rph xii) In *Science of Logic* is explicitly asserted that "State's reality is self-conscious individuals." (WL II 410)

Regarding the State itself, I cannot see how we could doubt it is the set of duties and right that entwine individuals. Especially when it is obvious that a State remains existing even when a monarch vanishes and another rises up, or even when a government vanishes and another rises up. Hegel just offers a definition of State that corresponds with the facts.

Previously, on the third part, we saw that it is not strength or positive Right's sanctions what maintains a State in existence. There would not be strength or police enough if, as it were, 'some fine morning' the whole people decided to disobey. As Napoleon said, bayonets are good for many purposes but for sitting on them. In the transcribed paragraph (Rph no. 280 Z) is obvious that the word law does not refer to positive Right as such; the objective part of State, in contrast with the monarch's subjectivity, is the true Right with the consequent articulation and structuring. As Hegel remarks in the corresponding section of the *Encyclopedia*: "Right should not be taken in the narrow sense of a legalist Right, but as covering *every* content of freedom" (EPW 486).

In his legal philosophy Hegel held the following perfect formula: "the commandment of Right is: be a person and respect others as persons." (Rph no. 36) This is clearly not positive Right. Given that the aforementioned is the true content of Right, Hegel held this decisive thesis for science: "the State lies upon thought, its existence depends on men's mentality; it is a spiritual realm, not a physical realm; spirit is essential." (GP I 507)

Government is a natural person or a group of natural persons. In their materiality, the sanctions applicable by the government are physical deeds: imprisonment, death penalty, fine, etc. On the contrary, the fact that certain set of human beings constitutes a State is not a physical datum; there is no way in which the State could be empirically verifiable. Physical presence at certain territory evidently is not the same that belonging to the State, since an individual does not stop being part of his State if he travels abroad. On the other hand, people that do not form part of a State can be physically present in its territory. It is not empirical either that the government's physicalities and its sanctions constitute Right and have to do with the State; empirically they cannot be distinguished from the violence executed by a shear of criminals sufficiently strong and organized. The conformity of such physicalities with the constitution does not make them empirically Right and State, since the fact itself that the constitution is Right is not a physical or empirical datum. Both State and Right are ideas.

"The spirit is just the State *in consciousness,* just so far as it considers itself as object." (Rph no 258 Z)

"The idea touches ground on the State the moment it acquires existence and reality *in knowing and willing*" (Rph no. 270 Z) (Italics added).

Rousseau had already said it: "Deep down, the political body, being only a legal person, is nothing but a reason entity." (1964, 608)

Since rulers are physical objects while the State is a group of ideas, the distinction between State and government is obvious. But Hegel goes further: he attributes governments of well structured States so little importance as he did to the monarch (head of State). The government is an instrument of the State, but in a well articulated State the important decisions are already made by the time they reach the governmental level, so that the instrument is in charge just of executing them. "Governmental affairs are of an *objective* nature, already substantially decided, and it is duty of some *individuals* to carry out and realize them" (Rph no. 291). Hegel would concede, as he concedes regarding the monarch, that there are underdeveloped stages in which the government plays a preponderant role, "but then we deal with a non-fully developed State, which is not well built".

> States' robustness properly resides in the communities. Government comes across with legitimate interests that it ought to respect; in so far the administration can only favor them but also custody them, the individual finds protection for the exercise of his rights and this is how his particular interest of conserving the whole arises. Recently the main efforts have been for organizing from above, but the lower parts, the massive about the whole remains somewhat inorganic; nevertheless, it is of supreme importance that it becomes organic, since only then it turns into strength and power. Otherwise it is just a heap, a multitude of atoms. There is legitimate strength only at the organized condition of particular spheres. (Rph no. 290 Z)

Only this is decisive. Hegel, it is true, adds there that power so characterized is monarchical. But I don't know how someone can be scandalized when, as we have seen, Hegel explicitly states the little importance of the monarch and the little importance of the government. He even adds right there that it should be distinguished between the antique monarchies and modern monarchies which are characterized by the autonomy that the particular spheres have from it. "A first monarchy must be distinguished from a second one." (VG 147)

The modern one is not juxtaposed on equality with democracy and oligarchy as a third form of government; that antique division was based upon quantitative criteria: it asked if it were one or few who governed. On the contrary, modern monarchy *includes* democracy: Rph no. 273 A. This is one of the most substantive theories that can be posed in philosophy of history.

It actually includes it, making that for the first time in history it is true democracy. Political thought cannot ignore this fact any longer: the Greeks invented the *word* 'democracy', but *democracy* is a modern European invention, it is a quite recent discovery in human history. Among the Greeks, four fifths of the population were slave, and that cannot be called democracy. Besides, the strong decisions were made by the oracle, not the people: "That democracy did not have yet the strength and energy of self-consciousness, which consists in the fact that *it is the people themselves who decide*." (PR II, II 189) (my emphasis) Whoever reads this expression cannot doubt about Hegel's democratic sincerity. It is related with what we were reading just a moment ago: "There is legitimate strength only at the organized condition of particular spheres (Rph no. 290 Z).

Whether we agree or not with monarchic *form*, I cannot see how one can doubt that Hegel is right when he says that modern European monarchy includes democracy: England, Sweden, Spain, Belgium, the Netherlands, Norway and Denmark are monarchies and they certainly are among the ten or twelve most democratic countries in the world. Hegel's critics would have to demonstrate that here are more democratic nations than these. Even an opponent as dumb as Findlay is forced to recognize:

> Despite Hegel's strange belief in hereditary Monarchy as the crowning truth of the State Idea, his view of the Monarch's functions are far from feudal, and are, in fact, in accord with modern British constitutional practice. The Monarch is merely the necessary apex of the State-structure, and as such he is merely someone who dots the i's, and whose individual character is not of great importance. (1958, 325)

Regarding this latter claim it might be suitable to recover the testimony of a mere historian, David Harris: "In England it came to be of relatively little importance who wore the crown." (EB 18, 745, 1)

Hegel insists in the little importance of what his critics, turning a deaf ear, consider central of his doctrine: "When there are firm laws and the organization of the State is well determined, the only thing that is left to decide to the monarch is, in comparison of substantial issues, rather unimportant." (WG 937).

What is central within the Hegelian conception is people's organization, the organic structuring of the communities and spheres of activity, which allows "to decide the people themselves", as Hegel tells us in contrast with the Greeks. This is why Hegel attributes so much relevance to the parliament:

> This opportunity of knowledge has the universal quality that only through it public opinion is capable of reaching real thoughts, of understanding the situation and concept of State and its affairs; and along with it, the capacity to judge it in a more rational fashion (Rph no. 315).

And, with regard to public opinion:

> Public opinion had great strength in all times and foremost in our time when the principle of the subject's freedom has so much importance and significance. Whatever is in vigor today is not in virtue of violence and much less in virtue of custom and habit, but in virtue of intellection and reasons (Rph no. 316 Z).

We have seen that the State really distinguishes from government and hence, those who accuse Hegel of propitiating authoritarianism have understood nothing. And we have seen that, despite the consistent appearances in the monarchic *form*, the decisive part of the democratic principle is present not only in the Hegelian system but it lucidly concretes in what we call today self-management. And it is, as sooner or later both left and right wings would have to face, the only way in which the democratic principle could be thoroughly realized.

All the misconceptions of the Hegelian political thought have just been useful impediments for the reception of a message of enormous importance precisely about democracy. A message capable of revolutionizing today's whole political thought. In order to deliver this message let us ask a crucial question.

If 80% of the population voted for the other 20% to become slaves, would all the supporters of democracy reject such decision? Why? Will it be congruous?

Evidently the way out claiming that the constitution prohibits it would be useless. Every constitution is modifiable by a supermajority and 80% is a supermajority. It would be enough voting first for a constitution modification.

This possibility is not unreal: remember that African-Americans in the United States are not even the 20% of the whole population. Other countries might think of persons having certain physical traits known to be possessed by less of the 20%: size, eyes color, blood type, birth weight, skin color, etc.

I repeat the question: If democracy consists on voting, with what logic could democrats deny the majority the faculty of deciding that certain minority should become slave?

This question shows that Hegel is right when he holds that the essence of democracy need not identify with the republican *form*, which is voting. Let it be clear that this book´s author prefers, against Hegel, the republican form, but here we are analyzing this issue under strict logic. And at this level Hegel's following paragraph touches the heart of the matter:

> From the viewpoint of the superior principle it becomes a subordinate and indifferent discrepancy what is usually considered as essential to a constitution, namely, if the individuals give their subjective acquiescence or not. It should first be determined if individuals are conceived as persons, if substantiality as spirit is present or not, i.e., as an essence known by them. (VG 145) (my emphasis).

This is why Hegel told us, as we saw above, that in a real State really important decisions are already made.

If every individual is conceived as a person, neither the majority nor anyone can treat someone as a thing. The first thing we should say, regarding our crucial question, is that *in a real democracy not everything is subject of vote*. There are a lot of things, precisely the most important ones, which we cannot leave to majorities, neither regarding a minority neither a single individual or in relation to anything. Deep down, this is the truth: *none of the important things are subject to voting*.

It is absurd to forget universal history when speaking about the State. Those who believe that man is man by nature and not by history forget that democracy did not exist from the beginning, neither did the Greeks who invented it. It was born in a time and in a civilization where all the people, including intellectuals and rulers, were iusnaturalists, i.e., *it knew* that the criteria for good and bad does not depend on any ruler or legislation or voting. Hence, democracy was born convinced of the fact that nothing really important is subject to vote. The democratic principle survives only within the framework of iusnaturalism due to both history and strict logic, as we shall see.

> In order to avoid slavery it must be known that *man as such* is free. But for that is required that man can be thought as a universal, without the particularity of being citizen from this or that State. The conception that man in general, as universal, is free, was not achieved by Socrates, Plato or Aristotle (WG 611).

"The situation is different at the European States; there the conception is general." (VG 145)

"Nowadays there cannot be legislators; legal institutions and judiciary relations are always present in our time. There is so little to add; just ulterior determinations of quite insignificant details can be provided by a legislator or a legislative assembly" (GP I 182).

If the aforementioned historical facts are forgotten, it is impossible to answer our question of reference. It is absolutely vital to realize this: *every* man is free, human beings are equal, and obviously *none* of this is empirical data. Therefore, this knowledge or conviction were not embedded in humanity from the beginning, they are not a natural endowment of the mind, and experience, for however bright we may suppose it to be, they cannot be acquired by the mind.

> That such equality exists, that is *man* and not just *some* men like in Greece and Rome, etc., who is recognized as a person with legal validity, is so far from being just by nature that, on the contrary, it is only a product and an outcome of the consciousness of the deepest principle of the Spirit and of the universality and development of consciousness (EPW no. 539 A).

That *every* man has infinite dignity, even though it is an absolute truth, by no means is an idea that humanity possessed from the beginning or that it could have acquired by nature. This is so because I would say, that is the most anti-empirical, affected and gothic idea that has ever been.

The civilizations that today have acquired it by contagion and universalisation from *a* civilization that discovered it. Even back in his time, Hegel claimed:

> Whole continents, Africa and the Orient, never had that idea and they still do not have it. The Greeks and the Romans, Plato and Aristotle and some Stoics did not have it. On the contrary, they just thought that man was really free depending on birth (as an Athenian citizen, Spartan, and so on) or by resolution of character or by education or by philosophy (the wise man is free even if he is made slave and gets shackled). Such idea was born by means of Christianity according to which the individual *as such* has an *infinite* value because he is object and end of God's love and is destined to have an absolute relationship with God as Spirit and to be inhabited by the Spirit, which means that *by essence* he is destined to supreme freedom. When within religion man acknowledges as his own essence the relationship with the Absolute Spirit, when entering the scope of *mundane existence* he also acknowledges that the divine Spirit is the substance of the State, family, etc. (EPW no. 482 A).

There is no historical objectivity in the political scientist that refuses to recognize that the idea of infinite dignity of *all* men is recent, considering that the human race exists since half a million years ago. There is no objectivity if he refuses to acknowledge that the idea has been spread recently from *one* region of the planet, exactly in the same way as anthropologists have tracked the historical local origin of certain discoveries that have become univesal, and they do not build up their hopes on some kind of spontaneous generation all over the Earth. The wheel, for example, was never discovered by Native Americans and they knew it by the diffusion which originated somewhere else.

Actually, the idea we are dealing with commenced existing for the first time in Europe, and its birth was due to the conviction that Jesus Christ, true God, had suffered and died for *all* men and it was *because* of this that it was discovered that everyone has infinite dignity.

> Subjectivity —regarding its infinite value— has suppressed every external difference, of dominion, of power, of social rank, even of sex and wealth. Before God all men are equal. This presents for the first time here and now unto consciousness, through the reflection and negativity of the infinite suffering of love. It is there that possibility resides, the root of a truly universal Right, which is the realization of freedom(PR III 178s).

It is, by the way, in this moment when the West discovered too that man by nature is not as he should be, since what man really *is* only in the Man-Christ had it been realized, in a total commitment for everyone's good. But there is nothing as forceful and revolutionary as the persuasion that one is not as one should be. This persuasion made the West unbearably upset with itself, it moved it in such way that the remainder cultures, in comparison with it are quite, static and inert. The West has set the whole world in motion.

We are now able to answer our old question. We said that not only historically but also logically, democracy supposes everyone's infinite dignity. The reason of being of democracy is the infinite dignity of everyone. Consequently, a democratic decision cannot condemn anyone to slavery. It would be a contradiction to itself and to undermine the only logical foundation that underlies it.

We are supposing, of course, that political thought pertains also to rationality. We are supposing that when we demand democracy and reject totalitarianism and dictatorship we do it rationally and not for aestheticism or whim or taste or strength, but because of reason. We are supposing that the demand of democracy has to *demonstrate to be with fundament* before those (and they really are out there) who prefer dictatorship and totalitarianism. It is not about oratory or poetry or strength, but of demonstration and objectivity. If the demand for democracy were not rationally *justified*, democracy and totalitarianism or slavery would be the same.

And well, the justification is that all human beings are subjects and not objects, which is a way of claiming that they have infinite dignity. Were not everyone to decide their own concerning issues, the few deciders would be treating the rest of the population as objects and not as subjects. Not only historically, but also logically democracy, presupposes a strict iusnaturalism.

How desirable it would be that the banal left-wing critique of 'formal democracy' at least understood it first considering its true consistency and its reasons, and only after that criticized knowing what it is about. If means of production are private property, the main *economic* decisions of the country are being made by a few individuals, who are treating the rest of the population as objects and not as subjects. Likewise, since there is no objective foundation (supply and demand is not) for assessing differently the different kinds of work that are necessary for society, it would be violating the equality and dignity to condemn the executers of certain kinds of work to an inferior way of life having them committed no crime. (It is another issue if I perform wrong my trade: that does deserve a sanction). One can even suspect intellectual dishonesty in the silence kept by liberals regarding the only possible justification of democracy, since that justification logically imposes conclusions incompatible with capitalism and with any classist society. On the other side, the revolutionary movement, if it is about rationality and not caprice, can not lay aside the infinite dignity of all human beings, for this is the only possible justification for their struggle.

Hegel denounces the incongruence of democrats that refuse to go to the rational bottom, to the ultimate reason of democracy's necessity: "Today we see the world full of the principle of freedom, and the latter specially related with the constitution of the State; these principles are true, but influenced by formalism they are mere prejudices because knowledge never gets to their ultimate reason" (PR I 309).

"The course that is in the grip of abstraction is liberalism, which is always beaten by the concrete and always goes bankrupt with the concrete" (WG 925).

> If it really is reason responsible of discovering these principles, it proves them so far as they are true and they do not remain purely formal, taking them to the knowledge of absolute truth, which is object only of philosophy. But Philosophy must get to the ultimate analysis, since if knowledge does not complete in itself, it is exposed to the unilaterality of formalism; but if it reaches the ultimate reason it reaches what can be considered supreme, it reaches God (PR I 309).

If liberals (and left-wings) realized how comical they look cautiously silencing the only possible justification of the demand of equality among men, sheltering at the irrational formalism that says 'we demand equality and democracy because we want to', political science will not be reduced nowadays to the enunciation of certain prejudices confronted with certain other prejudices.

Fortunately there is, as Hegel says, "the Right of the Absolute Spirit" (VG 147) and history laughs at the face of relativists and skeptics. The belief that *every* human being has infinite dignity has spread though the whole world. The essence of true civilization has won the battle against its rivals; it is still necessary to structure and articulate it in lots of places and in lots of aspects, but no one can doubt that we are heading towards it. "We should consider the Right of the Spirit of the world against the States." (VG 148). This is why Hegel told us that liberalism is in the grip of abstraction and the lack of content and always goes bankrupt with the concrete. The conviction of the infinite dignity of every human being, born in fact from the meditation of Jesus Christ's passion and death is an inalienable conquest of humanity.

The semi-anonym reviewer Z. C. (1822) of Hegel's legal philosophy considered inappropriate that in a treatise on Right as the realization of freedom Hegel meddled with universal history. With extreme naiveté, Z. C. asked: What does freedom care about universal history? (cf. Riedel ed. I 1975, 117)

On the margin notes Hegel points out afterwards, not without mockery: "Hugo is amazed that universal history is discussed along with the State." (Rph, Notiz zu no. 33)

In fact, only legal positivism and liberal formalism could come up with the simplicity of making static and timeless state science. There might be an underlying conviction, which I have already refuted (VI, 1), that man is free by nature and not by the tough work of history.

For Hegel there is no sovereignty that can stop the advance of the Spirit of the world. Regarding a savage tribe, a horde of barbarians that enslave and kill each other, he tells us: "their autonomy, as merely formal and lacking objective rights and firm rationality, is not sovereignty" (Rph no. 349).

It is a political and legal topic of enormous relevance and its intrinsic logic is the same that we saw regarding our question about democracy. Skepticism about the Hegelian Spirit of the world has worked here too as a pretext for neglecting an extremely actual and rationally irrefutable message. The existence of a State can only be justified if that State is the concrete realization of ethics and justice; the existence of a government is justified if it is indeed an instrument for that realization, since a man that rules over his fellow men evidently requires a justification. When international pressures demand that inside the country persons be treated as persons, what they demand is that the State be a State. It lacks of every rational support the government that gets shocked by those demands crying: interference! Its own authority does not exist if it does not consist in making that personas get treated like persons.

The principle of no intervention is positivist, and I demonstrated (VI, 3) that positivism is false. When someone says 'this is a State' or 'this is the government', *he does not make an empirical observation* but a moral judgment (true or false), since it is saying that certain set of men *ought* to proceed in such and such ways. But if the structure and the functioning of that certain set of men are violating human dignity, it is false that they *ought* to be obeyed and therefore it is false that there is State or government.

A government that systematically impedes the advancement of man towards justice and freedom is automatically illegitimate, is a shear of outlaws. History laughs in the face of positivists that claim that the world legitimacy has no meaning or that its meaning can be reduce to the possession of the necessary force in order to maintain certain 'order'. To affirm that there is order is a moral judgment, not an empirical observation. Hence, the positivist cannot reject the idea of legitimacy saying that it implies a moral judgment. Today everyone considers not only licit but mandatory pressing and interfering with the South African *apartheid*, despite that Peter Botha has the force to keep the population subjected. The UN General Assembly has just interfered nothing less than with the South African elections, rejecting almost by unanimity the results of municipal elections. Today everyone considers licit 'meddling' with the Chilean internal affairs and international pressures have forced in fact Pinochet to recognize his defeat by the supporters of the 'no'. If Hitler came back, the other countries would know they are not just authorized but obliged to interfere in defense of the German people. "The Right of the Absolute Spirit" (VG 147) demonstrates that it is over and above the Rights of nations, it scoffs at the anachronistic term called sovereignty.

International organizations just pretend to be positivists feigning that different governments actually "give their consent" for intervention due to the signing of international treatises. In fact, the health organization, for example, makes reports for all countries, even for those that have not sign anything and even reports on issues that are not mentioned in such treatises. Actually, smallpox has been almost eradicated despite the protests of some countries for the WHO's intervention in their territories. International organizations for education, food, childhood, work, and specially human rights 'intervene' all the same. Interferences exist, and its existence is good, and if only they were more vehement. That 'interference' does not have today a lot of means for self-affirming beyond condemnatory world opinion; this does not mean that it would not use them if it could.

The real danger of superpowers distorting Absolute Spirit in favor of imperialist ends contrary to world spread dignity and equality does not turn a true proposition into a false one. When we hear pain and panic cries at our neighbor's house we have the right and duty of getting in and stop a murder. The possibility that some perverted minds interpret and twist this right for other ends does not eliminate this right. If someone asks, what can guaranties that we do not get misinterpreted, we just answer that nobody. But rights and duties exist even if no one guaranties them.

It is blatantly obvious that it is not just for Mexicans the obligation to judge Mexico's affairs. Judging the truth or falsity of a proposition corresponds to any person endowed with reason and disposed for studying the issue with objectivity.

I suppose that other countries (perhaps not so many) have an equivalent for the Mexican Constitutional article number 33, second paragraph, according to which "foreigners could by no means interfere with the country's political issues". Jurists evidently have been too passionate when dealing with the rhetorical subject of sovereignty since they have not seen that every similar legal text is null and void for it is a *praeceptum de re impossibili*. The following question should be enough: what is exactly prohibited by such precept? Obviously that foreigners *influence* Mexican national political issues. So, even international news influence national politics, as public opinion polls and electoral outcomes demonstrates in developed countries. Hence, Hertzian waves should be prohibited over the national airspace or we should destroy every single radio and TV set along with the international section of every newspaper and magazine, as well as forbidding international travels both inbound and outbound. Particularly, it should be prohibited importing and translating books as well as the spreading of their ideas since anything influence politics more than ideas. Editors, translators and university teachers turn out to be infringing the aforementioned precept because they allow foreign thinkers to influence the political affairs of the country. The whole issue is ludicrous. If sovereignty lies on such precept, worst for it. Two centuries ago Hegel asserted that such a rhetorical vacuity was impotent against the advance of history's reason.

This is precisely why he held that "the rational is real" (Rph xix), i.e., that the rational ends up imposing sooner or later in the world, because "it is not as impotent as for ought to be and not being real." (EPW no. 6 A) I was said that the conviction in the fact that *every* human being has infinite dignity has been extended to the whole world and all peoples appeal essentially to it. So this is what is rational, the most rational that is and that it could be. "In contrast with all this, which is by itself and for itself the universal and substantial, everything else is subordinate and works as mere means." (VG 87)

The second half, "and the real is rational" (Rph xix) has provoked a sensational scandal among political thinkers and moralists. Against it they allude to alleged irrational empirical facts. Hegel replies: "Who is not cleaver enough to see that around him lots of things are not as they should be?" (EPW no. 6 A) The objectors believe that Hegel was some kind of fool unable to see what they see. On the contrary, they should wonder if they have actually understood Hegel's phrase. But "a sensible consideration of the world distinguishes, throughout the vast realm of interior and exterior existence, what is a mere apparent, transitory and insignificant *phenomenon*, from what in itself deserves to be called *reality*" (EPW no. 6).

The objectors overlook that it is about *universal history*. If the abstract intellect moors to its conceptual stiffness, even the flight of a fly is real: but since when a merely anecdotic, fortuitous and inconsequential event deserves to be called 'historic reality'? Beside all this, there is the fact that the objectors cannot define 'real', which is the most ludicrous thing of all this issue.

Hegel's renowned phrase also makes sense in the epistemological setting of the preceding chapters. True reality, from which the concept of 'real' is originally extracted, is reason, spirit (cf. III, 3); in this sense the rational is the more real that exists. On the other hand, it is just from the concepts of *reason* that we have access to reality; sensibility does not apprehends being. "No sooner man speaks; there is already a concept there" (GP I 336). In this sense, whatever is real for us is automatically rational.

A last word should be said in this fourth section. A good deal of historians of political ideas have been unable to understand the true common denominator of every iusnaturalist. Plato, e.g., is a iusnaturalist and does not appeal to nature but to eternal ideas of good and evil. Descartes is too a iusnaturalist and he does not appeal to nature but to innate ideas of good and evil. Also Suárez too, but his criterion is reason, not the whole human nature. Obviously, for iusnaturalism it should be understood the following: the fair or unfair, good or evil character of an action does not depend on any positive law. Every iusnaturalism shares this unique thesis; the name 'iusnaturalism' is completely accessory. The name produces equivocation, as if Right and morals have anything to do with nature, but it is too late to coin a new name since no one would pay attention to me.

Now, since Hegel's attack against natural is implacable, some superficial authors have believed that Hegel rejected iusnaturalism. They overlook that Hegel, while professing vehemently the content, he was aware of its name inadequacy and actually said where this inadequacy came from: "Those that cannot reach the universal take *what is by itself* to be something natural, just as they take the necessary elements of Spirit to be innate ideas." (GP II 107; my emphasis)

> The science of fundamental concepts of Right has been called natural Right, as if there was a Right that could concern man because of nature, and a Right, different from the latter, that originates in society; in the sense that society had to sacrifice natural Right as the true one. But in fact, [...] society is the suppression of the unilaterality of such principle and *the true realization of it* (NH 60s; my emphasis).

> The expression *natural Right*, which has been common within the philosophical doctrine of Right, leads to the ambiguity of whether Right exists *naturally and immediately* or if it is determined by the nature of the thing, i.e., by the *concept*. The former was the common usage and with it was created the fictitious *state of nature* in which natural Right would prevail. While, on the contrary, the condition of society and State would demand and bring with it a limitation of freedom and a sacrifice of natural rights. When the truth is that Right and all its decisions are founded only on free personhood, a *self-determination* which is rather the contrary of determination by nature (EPW no. 502 A).

It could not be said clearer that Hegel champions the true content of natural Right ("true realization of it"), but obliterating the equivocation of the expression; and also being quite aware of the fact that objective rights exist only in virtue that man is person, self-determination, which is achieved only thanks to intersubjectivity, society, as I have said (III, 7). It is about, Hegel says, *"laws and institutions that exist by themselves and for themselves"* (Rph no. 144). When I was talking about ethics I alluded to them and I said that they are innumerable, they are imperatives such as "You shall not murder", "you shall not deceive", etc., which are necessary for life in common.

> We have a system of ethical relations; these are duties and are within a system; every determination in place, subordinated one to other and the higher rules over the rest. This is how consciousness, which is higher than Stoic freedom, is obliged; determinations are secured to the Spirit; objective determinations called duties are kept as state of rights and they are valid in consciousness as firm determinations (GP II 295).

Those are the determinations that do not depend on any authority legislating or not, because the existence itself of society and persons requires them. The prototype of such contents is the small set of laws called the Decalogue: "The Universal Law was all the time the Ten Commandments," (Rph no. 216 Z) "the Ten Commandments which are the fundamental, universal, ethic and legal determinations of legislation and morality." (PR II, II 96)

Hegel highlights the iusnaturalist character of the innumerable set of duties and laws we are dealing with, i.e., its independence from any positive legislator, when discussing something about constitutions: "The mere existence of a people [...] presupposes a constitution, an organic condition, an ordered life of the people itself" (NH 530). This is why Hegel finds ludicrous the question that asks for the maker of a constitution: "It is easy to put forward the question: Who should make the constitution? It seems a clear question, but when we look closer, it shows immediately senseless. This is so because it presupposes that there is no constitution and that, hence, we only have an atomistic *collection* of individuals." (Rph no. 273 A) "Great laws, those properly significant, already exist; whatever is missing is insignificant" (WG 623).

"A constitution is not something merely done; is the effort of centuries and the idea and consciousness of the rational inasmuch as it has developed by a people" (Rph no. 274 Z).

Regarding the set of indispensable duties and institutions Hegel says: "Those institutions conforms the *constitution*, i.e., rationality developed and realized in the particular, and that is why they are sustain of the State" (Rph no. 265).

The content of iusnaturalism is present not only in the Hegelian system: it is the content of the system itself and it is devoid of the myth of natural man which made it vulnerable. If Hegel claims that the State is end in itself, it is because the State consists of the set of my neighbors: "The reality of State is the self-conscious individuals," (WL II 410) "the essence of the State is the alive ethical" (VG 112).

> Someone could say that the State is end in itself and that citizens are its means. But here the means-end relationship is absolutely inappropriate. The State is not something abstract, in front of the citizens. These are rather its constituents as in organic life, where no member is an end in itself and none is means either (VG 112).

"[...] individuals have something absolutely unsubordinated, something that is in itself eternal and divine [...] they are part of the end of reason itself, therefore they are ends in themselves [...] Man is an end in itself only due to the divine that is within him" (VG 106).

"Particular interest should not be ruled out and much least repressed, but harmonized with the universal, so that both it and the universal can be preserved" (Rph no. 261 A).

5. Family, society, state

Every confusion and debate about the relationship between civil society and State fade away once we consider that the State is *not* the government but "the State in itself and by itself is the set of the ethical," (Rph no. 258 Z), "the universe of the ethical" (Rph xxii).

In so far as they surpass animality and constitute a *human* phenomenon, i.e., an *ethical* phenomenon, family and civil society are part of such universe of the ethical which is the State. Just as morality and Right are true because of their ethicity and only that is the true morality and the true Right, so family and society are true because of the State and only in it they are true family and true society. "Every relation that due to form is private Right, are conceptually part of the State" (WG 917).

In civil society: "agents have in their activity finite ends, particular interests; but they also know and think. Therefore the content of their ends is pervaded by essential and universal determinations of Right, of duty, of the good. The mere desire, savagery and coarseness of will are left out of the stage and scope of universal history" (VG 95).

Even within the most basic economical relation —apparently pure egoism and the search of own benefits- there is and operating ethical relation: at least the ones required for communication through language (Cf. VI, 3), the one of not attacking, not fooling and 'keeping one's word'. A minimum degree of trust and gullibility is essentially required. Transactions are not between two savages carrying spears that with their free hand give and receive the object of trade. The day laborer is not paid after each stroke of their shovel; he has to trust and believe in his employer's word all day long. Workers are not paid every hour and not even every day; it is not give-and-take between belligerent parties. The entrepreneur believes too in the worker's implicit word that he wants to work and not to destroy the means of production. Without the ethical there would not be any collaboration or coexistence, without the State there would not be society.

Abstractly speaking, civil society is distinguished from the State, since "within civil society everyone is an end for his/herself and everything else is nothing" (Rph no. 182 Z), in civil society "all the waves of passion overflow" (*ibid.*).

Nevertheless, in *reality* civil society and the State are the same, since without the warp of the ethical, i.e., the State, without the entwinement of individuals by duties and rights, civil society would scatter, would stop existing, it would not be society. "Particularity gives the impression to subsist by itself, but it is supported and maintained by the whole" (Rph no. 270 Z).

This was showed perfectly clear by Plato (Rep II 361C; Laws VII 790B; 793B) and Aristotle almost twenty five centuries ago. Modern man should not be amazed that Hegel claims the same; thing are just like that. For example, Aristotle remarks: "In every association there is some Right;" (EthNic VIII, ix 1) from which follows that "it seems that every association are part of the State association." (*Ibid* 6) And it is what I was saying: as far as they constitute a human phenomenon and not a merely animal one, family and society are parts of the State.

Hegel puts it as follows: "In reality it is rather the *State* absolutely the first; only within it family develops in civil society; it is the idea itself of State which divides into these two elements" (Rph no. 256 A).

"The State is the *self-conscious* ethical substance—the unification of the principle of family and civil society" (EPW no. 535).

I have just mentioned why civil society, which is in the abstract a war of all against all and the mere quest of own benefits, in the concrete requires duties and rights to exist. This explains why Hegel claims: "Wherever there is civil society there is State [...]." (EPW no. 527 A). And also: "Civil society [...] presupposes the State, it needs it existence in order to survive" (Rph no. 182 Z).

Regarding family it might seem *prima facie* that such relation is not so obvious. But, if it is not an animal family, it is undeniable that among its members there are duties and rights and that it is in such ethical relation that the consistency of family as such really lies. I have already said (VI, 2 *fine*) that the relation between husband and wife is something quite superior to mere sexual instinct, and it is self-evident that if such tie were just sexual, the relation would not be stable nor conjugal society would properly exist. It is also evident that the duty of feeding, educating and looking after the children demands from the parents lots of things which cannot be in any sense reduced to the search of their own benefit. "It is through family that mans gets inside society, thanks to the reciprocal relation of social dependency, and this tie is ethical" (WG 888). Aristotle claims: "the State is essentially prior to household (*oikía*) and even to each one of us" (Pol I 1253a19).

> It does not matter that marriage has to do with love and sentiment; it is distinguished from the natural sensation of love, since it recognizes well-known duties independent from it even when love is gone. The concept and awareness of the substantiality of marital live [...] *constitutes the beginning of the State* as a realization of the rational and free will (Ästh II 496).

There is no need to insist more after the accurate analysis of Aristotle: in every human group a relation of justice takes part, and hence every group is part of the State. If this scandalizes some theoreticians it is because, infected with legal positivism, they believe it has something to do with the government; it is because they imagine that there are no duties and rights until they are in black and white and positively decreed by some authority.

In fact, "the State [...] is alive in so far as its two components —family and civil society— are developed within it. The laws that govern these components are institutions of rationality that shine in them" (Rph no. 263 Z).

"Life of civil society constitutes the ground of duty; individuals have their vocation pointed out, hence their duty is pointed out; their morality consists in behaving according to it" (VG 95).

Hegel does not admit that State is reduced to civil society (cf. Rph no. 182 Z), i.e., that a State is conceived following the *abstract* concept of civil society, whose content is the quest of one's own benefit and having each one as an end in itself. Plato and Aristotle already insisted in the fact that the State is *not* a means for the satisfaction of material necessities; the value of the State is not merely instrumental.

Such a conception of the State would mingle it with a private company or a military alliance. Holding that the State is a means would be pure and sheer immorality since the State is my fellowmen. Intersubjectivity and bonds among people is an end in itself, not means for something else:

> Unification as such is itself the true content and end, and the fate of individuals is to realize a universal life; this, which constitutes the substantially and universally valid, constitutes both the starting point and the upshot of any other particular desire or activity or behavior (Rph no. 258 A).

"The State is not one of those unions whimsically decided by individuals" (GP III 307).

Theoreticians like Hobbes, Locke and Rousseau explain the origin of the State by means of a contract, they never understood that the State *essentially* precedes man, but they never understood that the State makes the man, not vice versa. They never understood that by nature man would be animal and not man, they did not even understood that language, which is required for making the contract, is already a result of the organic intersubjectivity called State and it is even a constitutive part of the it.

"The highest duty of an individual is to be member of the State;" (Rph no. 258); "only from a spiritless perspective the State could be something merely finite" (Rph no. 270 Z).

"Today is known that the ethical and fair in the State is also the divine and God's commandment; and regarding its content there I nothing higher or holier" (WG 888). The State is an end.

Let us repeat it: the distinction between civil society and the State is abstract. In concrete reality civil society is identified with the State: "Wherever there is civil society there is State [...]" (EPW no. 527 A).

Now, it would be a mistake to believe that this depends on the definitions each one wants to come up with. The work of Plato, Aristotle and Hegel are *science*, not literature. They explain *how things are*.

On one hand, I showed (VI, 4) that the existence of the State does not depend on the existence of a government or on the existence of positive laws, but, on the contrary, it is presupposed for government and positive legislation can exist. Therefore, the State consists on the set of duties and rights that tie persons or, if one prefers, on a set of persons in so far as they are tied by duties and rights.

On the other hand, a mere cluster of individuals is not a society, and the intertwinement by mere economic relations or transactions is no way enough for constituting a society. For example, the inhabitants of Mexico's Northern borderline zone have greater economic relations with Americans than with Mexicans, but they still form part of Mexican civil society. Aristotle already defended this: Etruscans and Carthaginians which make continuous economic transactions among them would be one and the same society. (Pol III 1280a31ss). Material self-sufficiency is neither enough nor required in order to constitute a society, since nowadays almost no society is self-sufficient. Racial, linguistic or moral homogeneity cannot be imagined as the essential trait of a civil society or a State; it would be too easy to refute through facts. Territorial unity, whatever that means, cannot be postulated as an essential trait either; the existence of Pakistan refutes it. But, more importantly, one cannot make sense of the expression 'territorial unity': even when divided by the sea or the mountains, persons can form part of the same civil society. Mexico and the Philippines demonstrate it. Saying that people should not cross over a territory pertaining to another civil society would be a definition where the *definiendum* appears in the *definiens* making it a circular definition that defines nothing.

No single physical datum is useful for corroborating the existence of civil society. Therefore, it is impossible to define civil society except as a set of persons intertwined by specific duties and rights. And that is the definition of State.

6. Two issues about humanity

I will talk about a theological and an epistemological issue quite close related with what I have said in the present chapter as to leave them aside.

> It could be thought that the idea of salvation does not concern us [philosophers], because salvation is a future end, one of the afterlife. But then, the existence in this life would be just a preparation for that end. […] Individuals would have no alternative but to see as a mere means whatever leads them to salvation. But things are absolutely not like this, we should definitely rather conceive it as the absolute. Now, according to religion the end —both the natural existence and the spiritual activity— is God's glorification.

> And, indeed, that is the most praiseworthy end of spirit and history. [...] The end of spirit is to achieve consciousness of the Absolute, in such a way that this consciousness is the only truth and everything is arranged according to it, so that it is what reigns and has reigned in universal history. To know this as a matter of fact is what is called praise God or glorify the truth (VG 181s).

There cannot be two ends in the proper sense of the word. When the mind focuses on one, the other one turns into means. To call 'subordinated end' something that really is means would be a linguistic chicanery with no content, since subordination consists precisely in it being means. But a philosophy that goes deep into morality cannot accept that the State and universal history become means, because both the State and history *are* my fellowmen, the set of my fellowmen.

I explained (VI, 2) that the 'other' end, which theologians have invented despite it contradicts the Bible, is simply impossible.

Now, on that other end is based the notion of supernatural: supernatural is defined as whatever has to do with that other end.

If anything has been demonstrated in the previous pages is that man as man, precisely in as much as it differs from animals, has no nature. But then the idea of something supernatural for man is based on a false supposition: its content pretends to be something in contrast with man's nature, something that is over and above what is natural in man. But what is natural in man is not human but animal, man as such is always over and above and in contrast with what is natural, if not he would not be man.

We discussed (VI, 1) that naturalness is egoism, and sin consists in "deepening-on-self in nature", and every man commits that sin in his first free action. Every man needs irreplaceably divine assistance, divine imperative appeal, *for being man* and not animal. Abstract theology, on the contrary, makes the original sin to consist in the lack of a 'supernatural' grace which is not required *for being man*. Such conception in untenable, since it can only be called sin the lack of whatever man *ought to be*.

Such a theology, in the end, attributes to God a sin which pertains to man, since man is not liable for lacking such blessed supernatural grace. What really happens is that theologians *do not* accept the original sin; they do not admit that man is evil from the first moment. The Jahwist undoubtedly would say to them: Abandon all excuses and self-justifications, the truth is that *I* sinned of egoism from the first moment, *I* am the sinner; I deepened voluntarily my animality; it is not that such sin is attributed to me, but I rather personally committed it myself.

If it was not tragic for the historical consequences it has had, what happens with such theology would be funny: it turns out humanity needs Christ only to be able to convince God to stop attributing to us a sin that we did not commit but that God fancied to attribute to us.

The reader might allow me the following incidental remark that does not pretend to point out the cause of this theology: capitalism has been benefited by theologians' consideration that natural egoism is not sinful, the search of one's own benefit. Such egoism is what capitalism uses as the engine of the whole system, and justifies it based on its presumed lack of culpability.

In one of the deepest ever written paragraphs, Hegel calls pantheism the traditional denial of the original sin:

> If man is God immediately, i.e., if as an individual he considers himself God, that is pantheism. It is coherent with this the opinion that man is good by nature, that as an individual he is affirmative [...] This is why it has been said of such self that God is not within him nor he is within God, and that is has to do with God just in an extrinsic fashion (PR I 255).

In other words: if I am not evil, I am the true God; I do not need God for anything. If my self am self-sustained, that is, if my self-consciousness gets to exist without God's intervention, I am not a creature, I am God.

It is worth noticing that still during the fifth century the Church taught that grace and divine assistance are required for fulfilling the commandments, which means in order that man can truly be man. (See Denzinger 137-138). Under this conception, faithful to the Scripture, grace is not conducive to a 'supernatural' end.

Theology, like other particular and specialized disciplines, did not bother to criticize its own concepts and to ascertain if they have meaning or not (cf. II, 7 *initio*). Human nature would be the same as human essence, but as a principle of activity, not as a principle of being. It is an explicative concept, not a descriptive one. But I have already established (V, 4) that those explicative words, specially 'essence', are just tautologies and smokes and mirrors. Regarding man, however, the argument against an alleged explicative factor has more weight, since man's activities and actions would not be self-determined, they would not be free, but determined by human nature. If it is a free act, any other factor but itself can *explain* it. And if nature does not fulfill any explicative function, it does not fulfill any function and it is idle, it is a concept lacking content.

Someone might rebut: human nature influence, it does not determine. I answer: if it only influences, it is something exterior, since lots of thing influence, e.g., what we see, hear, feel, etc. What is natural in man is the animality, but man is man only in so far it is free from nature; such nature is not *man's* nature, but animal's. The concept of human nature is incompatible with free will, with the concept of spirit.

They might say that for human nature one should understand 'what is characteristic of man'. But what is characteristic of man is being spirit, and he shares that with God, which is precisely what theologians would like to contrast with man's nature and hence speak of the 'supernatural'. And the only meaning that such an adjective can really have is what cannot exist because of natural causes. But that is what man is, the spirit. "Only the spirit is the absolute interruption of nature; only it is the true miracle against the course of nature, what is really affirmative against it" (EGP 176).

Notice that the moral use of the expression 'human nature' is quite widespread and it has nothing to do with the ontological use or meaning that I have contested. Since they still do not understand that the imperative emerges from the infinite dignity of the neighbor and hence the criterion is evaluating if an action treats the neighbor as a subject or as an object, there are moralists that use as criterion of morality the conformity or not with the alleged human nature. Evidently there is Rousseau in the middle, since it is supposed that what is natural in man is good.

Even if we can do without every supposition, it is time to denounce the deep immorality of such a moral criterion, regardless of how widespread it is among moralists and theologians. There are natural tendencies towards incest, there is a natural tendency towards egoism, there is a natural tendency towards aggression, irresponsibility, sloth, cowardice, infidelity, etc. Founding morality in nature is one of the most resounding mistakes ever made. It is tantamount to affirming an obligation to resemble animals.

The epistemological question concerns all human sciences, historiography included. It would be interesting to know how people within these disciplines would react if told that their object of study, man, is not an empirical datum. Most of the time, of course, we immediately intuit if the object that we have in front of us is a human being or not. But this is not the question; the real question is: How do we intuit it? What kind of knowledge is it? Is that intuition an empirical perception or rather intellectual knowledge?

It is not a fact perceived through the senses that the body in front of us possesses thought. As Hegel remarks, "only the spirit perceives the spirit" (EGP 176). We know that the object is a human being only when we acknowledge its capacity to think. "Everything that is human is such only in so far there is operating thought in there, and outer look can be anyhow; it is human only to the extent it possesses thought" (EGP 81).

> In everything human the active part is thinking, the thought. The animal lives too, just like man has needs, feelings, etc. If there is a distinction between man and animal, the feeling has to be human, not animal, i.e., there has to be a thought involved [...]. In close inspection, we conclude that thinking is not a particular trait, some special capacity, it rather is what is essential, the universal that produces everything else (EGP 82).

A blind person is able to distinguish when he is dealing with a human and when not. Hence, it is not visual data which allow us to identify a man. A deaf personal is able too to make the distinction. Hence, it is not sound which make us confirm the existence of a man.

It might be convenient to recall the testimony of a good anthropologist, Leslie White: "It is the symbol which transforms an infant of Homo sapiens into a human being; deaf mutes who grow up without the use of symbols are not *human* beings." (1964, 41) "A baby is not a human until he begins to symbol" (ibid. 52). The existence of a symbol is something empirically indemonstrable. One can *see* certain physical data, *hear* certain sound, but we do not see, nor listen that such material thing *represents* something; it is a relation that we can only grasp with our understanding. Therefore, the fact that the body we have in front of us is using symbols is not an empirical datum but fact perceived by reason. It is only then when we know that the body is a human being.

It amazes me that anthropologists do not even seem to suspect what the epistemological implications are of the procedure by which —based on common sense certainties— (they think) they learn *a posteriori* what is man. According to them, every behavior, configuration and institution man has or have had constitutes an *a posterior* datum. It is from this datum that, by whole induction, one should figure out what is man. But, how do they know which behaviors are human and which are not. According to which criterion should they be included or excluded from the set? They will say: it is enough that man has them. But how can they determine that is man an organism with certain behavior? Evidently because before the experimental inquiry they already possess a criterion for determining which bodies are men and which not. Hence, it is absolutely false that experimental inquiry teaches them what man is, since they knew it beforehand.

In the end, saying 'this is human' is making a huge value judgment. It is the anti-moralist allergy what impedes human sciences from recognizing that the starting point of all of them is a moral judgment and makes them to blindly hold the thesis that they are empirical disciplines.

Before throwing a materialist thesis about man (e.g. that he differs from animals only in degree, or that the world psyche is just a name, just like 'heat', etc.), the psychologist or the anthropologist has to offer a justification for the fact that he actually uses the word man. It is not a universal obtained by generalization of *a posteriori* data, since it is required for building up the set of objects from which they pretended to obtain it by generalization. They have no right to use that word if they have not known a being specifically distinct, irreducible.

In order to keep the illusion of empiricism, it is common to believe that the lack of facial hair is enough for distinguishing a man from a monkey. But in 1986 there was a Mexican TV show that presented a family from Loreto, Zacatecas, whose children have their entire body covered with hair —including forehead and head— it must be said, with hair longer than a simian. Scientists from all over the world came to confirm the fact, and anyone can visit today that family. The children are normal in everything else: they speak and think as any of their neighbors.

It is usually also mentioned, with great vagueness, 'human appearance'. But Harlan Lane quotes John Locke who speaks of certain priest called Saint-Martin who, when he was born,

> ...had so little the figure of a man that it bespake him rather a monster. It was for some time under deliberation whether he should be baptized

> or not. However, he was baptized and declared a man provisionally, till time should show what he would prove. (1979, 31)

Who says 'this is a man' is saying how he *ought* to treat the body in front of him. It is a moral judgment.

In the second half of the eighteenth century something happened to European thinkers that should be enough to demonstrate the anti-empiricism that I have been posing. Lane himself sums up:

> The search for terms of comparison led out of society and into near and distant wilds. Numerous expeditions returned with samples of alien flora and fauna, including a parade of orangutans, gorillas and chimps as well as Pygmies and Hottentots, all of whom were subjected to detailed naturalistic and anatomical observation. Orangutans (whose name comes from Malaysian and means *homme sauvage*) were especially prized. [...] Many of the finest minds in Europe were busy dissecting the same orangutans, and even proposed, to the outrage of the cergy, that one be mated with a prostitute. (1979, 20)

It is clear that the lack of hair or tail did not seem to Europeans, neither to the Malaysians, a sufficient empirical datum for identifying a man.

The verticality of the trunk was considered too, for long time, a safe sign. But J. R. Napier, from London University, specialist in primate biology, confirms: "The view that the possession of uprightness is a solely human attribute is untenable; man is merely one species among the 189 that constitute the order who has exploited the potential of his ancestry to his ineffable advantage" (EB 23, 425, 2).

There were some physicalists hopes placed in the opposable thumb, but that trait has turned out to be particularly vulnerable, since some marsupials have it, while some primate lack it.

Regarding the brain's size, the differences among men are greater than those between the less gifted men and gorillas. There are men with 2000cm^3 and there are even some men with 850cm^3, while in gorillas there are cases of 685cm^3. If a difference of 165cm^3 were enough for settling the difference of species, human race would not be *one* species. Leslie White has to recognize: "So far as is known the only difference between the brain of man and the brain of an ape is a quantitative one", and he adduces the following to A. J. Carlson, which is very relevant: "man does not have new kinds of cells or new connections among those cells" (White 1964, 49).

It is interesting to notice that anthropoid monkeys have in the larynx and its neighborhood the whole required mechanism for producing articulated sounds. They do not speak, and there lies the whole problem; but physically they have everything is required to talk.

Many physicalist hopes were placed on our set of teeth, on whether the superior premolar has three or two roots, on the number and position of the cusps and grooves, on whether the size of the molars increases or decreases. But these are also differences of degree, and specialists talk with caution: 'generally' there is such and such in man. For example:

> The upper first bicuspid generally has two roots, and sometimes one. The second generally has one root, sometimes two; the inferior premolars generally have one root each. […] the upper molars have four (sometimes three) cusps whereas the lowers have five (sometimes four). The number and position of the cups and the character of the grooves between them differ among individuals. Generally, the upper molars have three roots and the lowers two. In persons of European origin, the first molar usually is large, the second smaller, and the third the smallest. In some other groups the third molar es larger than the second though smaller than the first (EB 17, 283, 1).

The reader understands that, with these premises, if afterwards some differences of this kind are discovered between the dentition of some monkeys and the human dentition, saying that *that* difference does mean a transition of species while the difference between the Tibetan molars and the European ones is not decisive is just an emotive arbitrariness. Evidently, whoever makes this decision already knew that the Tibetan and the European were men and the monkey was not. Such knowledge is not obtained from the analysis of the molars. This explains why the eminent English paleontologist Kenneth Page Oakley concluded that no physical evidence is enough to distinguish man from animal. Paul Overhage says: "Oakley explicitly empathizes that after the discovery of new fossils a definition of man is impossible on the base of just bodily and anatomical traits that clearly distinguish man from animal and from its bodily structure" (1961, 117).

In Hegel's time the shape of the teeth was as arbitrary as the lobe of the ear; so that Johann Friedrich Blumenbach, founder of the physical anthropology, pointed it out as an anatomical characteristic that distinguished man from the rest of the animals. It was discovered afterwards that some monkeys have it. And regarding all these physical issues Hegel said very well:

> By its exteriority it is easy to notice that the knowledge of the concept did not start in them; it was rather an obscure presentiment, an indeterminate but deeper sense, a guess of the essential, what preceded the discovery of species in nature and spirit; and only after that it was seek, for the abstract intellect, some determinate exteriority (WL II 456).

Up to our days, anthropologists and paleontologists have failed with all the physical traits they have proposed as distinctive, in spite of the fact that they ignored a fundamental methodological requirement: they were supposed to discover what was essential, i.e., a trait that explained the existence of everything else and of the set as such; it would have to be a trait that explain why man has thought and animals do not. This is way things like the ear's lobe are ludicrous.

Even if they were able to detect in the future a physical trait exclusive of man, Hegel's paragraph hit the bull's-eye: they first determine by other methods which existing bodies are men and which are animals, and only after that they look for a physical trait that adorns the former and is absent in the latter. This means that the concept of man was not obtained *a posteriori* but through self-consciousness. When we say 'this is a man' we actually say 'he is like me'.

It is self-determination what we perceive through self-consciousness as constitutive of the self. "I am only what I am related with my freedom" (Rph no. 117 Z).

When Ortega says 'I am myself and my circumstances', he is no doubt thinking in the influence that other factors, including animality, can exert over me. But, if that was the case, things like air, soil, vegetation, the Sun, and the whole cosmos should be incorporated to the definition of man. But then the definition does not accomplish its goal, the essential goal of every definition, which in this case would be distinguishing man from everything else. It lacks rigor such insolence from Ortega; it overlooks that the mental operation called definition is subject to certain essential rules. Hegel said that man is not free because certain movement commences within him, but because he can stop it and decide himself to what point such factor is determinant or not at all. This is why I am only what am related with my freedom.

When I call other objects men, the only thing I mean is that they are like me. It is principally a moral act. Referring to the aforementioned monstrous priest, Harlan Lane comments:

> But savages, primates and wild children all share with the abbe this lack of resemblance to the man in the street. The status of savages was decided once and for all by papal bull in 1537: the Americans discovered by Columbus were declared human. (1979, 22)

There were no empirical data by which the Europeans could know if the Indian and the black are men or not. The atrocity committed by some or many Europeans when denying them their humanity did not consist in that they were *seeing* that they were men and they denied it knowingly; but in the fact that they did not executed the *moral* act of recognizing that the other *is like me* and hence he is an end and not means and hence he has infinite dignity. Without this moral act is impossible to corroborate that the object in front of me is a man. You do not know that such object is a man as long as you do not perceive the 'you shall not murder'.

"What is free exists only for what is free; only for a free man the other exists as free too" (PR III 94).

"The kind of relationship of man with other men, that is what is human" (PR II, II 65).

Those who think that demonstrating the infinite dignity of every human being is hard, is because they think that *first* they could verify empirically the existence of a human being, and *afterwards* it should be inquired if he has or not the property of infinite dignity, which, for being a valuation, lies outside the scientific. But 'man' is not empirical data. There is no knowledge of a man without moral obligation regarding that man, i.e., without categorical imperative. And this is synonymous with infinite dignity. The concept of 'end and not means' has no other content than the one of infinite dignity. "The Scripture says that God made man in his image; that is the concept of man" (PR III 127).

Historians and philosophers fool themselves in the same fashion when they believe they can constitute without value judgments the object of study of historiography and this is why they reject that human history has a goal and is directed toward an end. They think that there are many 'real' facts that are not directed toward the thorough realization of the concept of spirit. But how do they know that such facts are *human* and hence that they pertain to history? "The mere desire, savagery and coarseness of will are left out of the stage and scope of universal history" (VG 95).

Human acts are directed towards the humane, towards what is every time more humane. Without such finality they would not be human. Humanity is not an instantaneous entity; it is not a timeless verifiable entity. Its movement towards true humanity pertains essentially to the object. Loose facts and particular cultures can only be judged and *described* according to this movement. Those who try to verify humanity as it is at a particular moment fool themselves trying to do without an end ofhistory. They are implicitly saying that it is completely human as it is. This is a value judgment, false by the way.

Bibliography

Adorno Theodor W., *Tres estudios sobre Hegel,* tr. Víctor Sánchez de Zavala, Madrid, Taurus, 1969.
Althaus Paul, *DerBriefan die Rómer,* Góttlngen, Vandenhoeck, 1966.
Baker J. J. W. y G. A. Allien, *Biología e investigación científica,* tr. George y Figueroa, México, FEI, 1970.
Bastin Ted (ed.), *Quantum Theory and Beyond,* Cambridge, U. R, 1971.
BlokhintsevD. I., *The Phiiosophy of Quantum* Mechanics, Reidel.Dordrecht, 1968.
Bohr Niels, Física *atómica y conocimiento humano,* tr. Albino Yusta, Madrid, Aguilar, 1964.
Bonsiepen Wolfgang, cf. Póggeler.
Born Max, *Einstein's Theory of Relativity,* Nueva York, Dover, 1962.
Brillouin León, *Relativity Reexamined,* Nueva York, Academic Press, 1970.
Bunge Mario, *Philosophy of Physics,* Dordrecht, Reidel, 1973.
Capek Milic, *El Impacto Filosóñco de la Física Contemporánea,* tr. Eduardo Gallardo Ruiz, Madrid, Tecnos, 1973.
_____, *Bergson and Modem Physics,* Dordrecht, Reidel, 1971.
Carnap Rudolf, *The Logical Structure of the World and Pseudoproblems in Philosophy,* tr. Rolf A. George, Berkeley, University of California Press, 1969.
Cranfield C. E. B., *The Gospel According to St.* Mark, U. P., Cambridge, 1966.
D'Abro A., *The Evolution of Scientiñc Thought from Newton to Einstein,* Nueva York, Dover, 1950.
_____, *The Rise of Modern Physics,* 2 vols., Nueva York, Dover, 1952.
Dingle Herbert, cf. Schilpp.

Dirac P. A. M., *The Principies of Quantum Mechanics,* Oxford, Clarendon, 1981.
Dyson Freeman, *Disturbing the Universe.* Harper and Row, Nueva York, 1979.
Eddington Arthur, *Space-Time and Gravitation.* U. R. Cambridge, 1978 (1920).
Einstein Albert, *La Relatividad,* tr. Ute Schmidt de Cepeda, México, Grijalbo, 1971.
———, y Leopold Infeld, *La Física, aventura del pensamiento* (orig. *The Evolution of Physics), ti.* Rafael Grinfeld, Buenos Aires, Losada, 1984.
———, cf. Schilpp i y n.
Elias Norbert, *Über den Prozess der Zivilisation,* Frankfurt, Suhrkamp, 1977, 2 vols.
D'Espagnat Bernard, *Conceptual Foundations of Quantum Mechanics,* Massachusets, Benjamín, 1976.
Evans-Pritchard E. E., *Social Anthropology,* Free Press, Nueva York, 1951.
Findlay J. N., *Hegel, a Re-examination,* Londres, Alien Unwin, 1958.
Frisch O. R, cf. Bastin.
Garaudy Roger, *Dieu est mort, étude sur Hegel* París, PUF, 1962.
Grundmann Waler, Das *Evangelium nach Markus,* Berlín, Evangelische Verlagsanstalt, 1968.
Gunkel Hermann, *Génesis,* Góttingen, Vandenhoeck, 1966.
Harris Marvin, *The Rise of Anthropological Theory,* Nueva York, Crowell, 1968.
Heinrichs Johannes, *Die Logik der 'Phánomenologie des Geistes',* Bonn, Bouvier, 1974.
Heisenberg Werner, "Über quantentheoretische Umdeutung kinematischer und mechanischer Beziehungen", *Zeitschríñ für Physik 33* (1925). pp. 879-893.
Henrich Dieter, *Hegel im Kontext,* Frankfurt, Suhrkamp, 1971.
Hill Winfred, *Teorías contemporáneas del aprendizaje, tr.* Eva Flora W. de Setaro, México, Paidós, 1983.
D'Hondt Jacques, *Hegel, filósofo de la historia viviente, tr.* Aníbal C. Leal, Buenos Aires, Amorrortu, 1971.
Hyppolite Jean, *Genese et structure de la phénoménologie de 1 'esprit de Hegel,* París, Aubier, 1964, 2 vols.
———, *Logique et existence,* París, PUF, 1953.
Jammer Max, *The Conceptual Development of Quantum Mechanics,* Nueva York, McGraw-Hill, 1966.
Jiménez Huerta Mariano, *Derecho penal mexicano,* México, Porrúa, 1968ss, 4 vols.
Kaufmann Walter, *Hegel, a Reinterpretation,* Nueva York, Anchor, 1966.
Kelsen Hans, *Teoría pura del derecho,* tr. Moisés Nilve, Buenos Aires, Eudeba, 1968.
———, Teoría *general del estado,* tr. Luis Legaz Lacambra, México, Editora Nacional, 1965.
Kilmister C.W., cf. Bastin.
Kojéve Alexandre, *Introduction á la lecture de Hegel,* París, Gallimard, 1947.
König Franz, *Cristo y las religiones de la tierra,* tr. Ramón Valdés del Toro, Madrid, BAC, 1960, 3 vols.
Koppers Wilhelm, cf. König.
Kroeber A. L., *The Nature of Culture,* Chicago, U. R, 1952.

Kroeber A. L., *El estilo y la evolución de la cultura,* tr. Ángel Guillen, Guadarrama, Madrid, 1969.

———, *Review of' Primitive Society'.* American Anthropologist 22 (1920), pp. 377-381

Kuss Otto, *Der Rómer-Brief,* Regensburg, Pustet, 1963.

Labarriére Pierre-Jean, *Structures et mouvement dialectique dans la phénoménologie de l'Esprit de Hegel,* París, Aubier, 1968.

Lane Harlan, *El niño salvaje de Aveyron,* tr. Mario Carretero, Madrid, Alianza, 1984.

Von Laue Max, cf. Schilpp.

Lauer Quentin, *Essays in Hegelian Dialectic,* Nueva York, Fordham University Press. 1977.

———, *A Reading ofHegel's Phenomenology of Spirit,* Nueva York, Fordham University Press, 1982.

———, *Hegel's Concept of God,* Albany, State University of Nueva York Press, 1982.

———, *Hegel's Idea of Philosophy,* Nueva York, Forham University Press, 1983.

Lowie Robert, *Primitive Society,* Nueva York, Boni and Liveright, 1920.

Luhmann Niklas y Habermas Jürgen, *Theorie der Gesellschaft oder Sozialtechnologie,* Frankfurt, Suhrkamp, 1976.

Lukács Georg, *Der junge Hegel* Frankfurt, Suhrkamp, 1973, 2 vols.

Maier Josef, *On Hegel's Critique of Kant,* Nueva York, AMS, 1966.

Marcuse Herbert, *Razón y revolución,* tr. Julieta Fombona de Sucre, Madrid, Alianza, 1972.

Margenau Henry, *Physics and Philosophy,* Dordrecht, Reidel, 1978.

Maslow Abraham H., *Motivation and Personality,* Nueva York, Harper and Row, 1970.

———, The *Farther Reaches of Human Nature,* Nueva York, Viking, 1975.

Michel Otto, *DerBrief an die Römer,* Göttingen, Vandenhoeck, 1966.

Miller Gary J., cf. Rossi.

Mure G. R. G., *An Introduciion to Hegel,* Oxford, U. P, 1966.

———, *The Philosophy of Hegel,* Oxford, U. P, 1965.

Von Neumann Johann, *Mathematische Grundlagen der Quantenmechanik,* Berlín, Springer, 1932.

Newton Isaac, *Mathematical Principles of Natural Philosophy, Optics,* Chicago, Encyclopaedia Britannica, 1977.

Nineham D. E., Saint *Mark.* Penguin, Londres, 1964.

Overhage Paul y Rahner Kart, Das *Problem der Hominisation,* Freiburg, Herder, 1961.

Parry Clive, cf. Sorensen.

Pöggeler Otto ed., *Hegel,* München, Alber, 1977.

Poincaré Henri, La *Cience et l'hypothese,* París, 1909.

Popper Karl, *The Open Society and its Enemies,* Londres, Routledge, 1974, 2 vols.

Popper Karl, *Objective Knowledge,* Oxford, U. P. 1973.

Post H. R., cf. Bastin.

Rahner Karl, *Geist in Welt,* München, Kösel, 1964.
______, *Schriften zur Theologie,* 13 vols., Zürich, Benzinger, 1954-1975.
Redfield Robert, *The Little Community* and *Peasant Society and Culture,* Chicago, V.P. 1960.
______, cf. Rossi.
Resnick Robert, *Introducción a la teoría especial de la relatividad,* tr. Ricardo Luna Victoria, México, Limusa, 1981.
Riedel Manfred ed., *Materialien zu Hegels Rechtsphilosophie,* Frankfurt, Suhrkamp, 1975, 2 vols.
Ripalda José María, *La nación dividida,* México, FCE,1980.
Ritter Joachim, cf. Riedel.
Rosenblueth Arturo, *El método científico,* México, Prensa Médica, 1984.
Ross Herbert H., *Biológica! Systematics,* U. S., Addison, 1974.
Rossi Ino *et al.,* eds., *Anthropology Full Circle,* Nueva York, Holt Rinehart Winston, 1977.
Rothstein Jerome, cf. Bastin.
Rousseau Jean-Jacques, *Oeuvres complétes III* (Pléiade), París, Gallimard, 1964.
______, *Emilio.*
Schebesta Paul, cf. König.
Scheibe Erhard, *The Logical Analysis of Quantum Mechanics,* tr. J. B. Sykes, Oxford, Pergamon, 1973.
Schilpp Paul A. ed., *Albert Einstein, Philosopher-Scientist.* Open Court, La Salle i, 11,1969, 2 vols.
Schrödinger Erwin, *Über die Unanwendbarkeit der Geometrie im Kleinen,* Naturwissenschaften 22 (1934).
Schubarth K. E., cf. Riedel.
Schweizer Eduard, Das *Evangelium nach Markus,* Göttingen, Vandenhoeck, 1967.
Serway R. A., *Física,* tr. José Pérez Castellanos y Adolfo Bustamante, México, Interamericana, 1985.
Sorensen Max ed.. *Manual de derecho internacional público,* FCE, México, 1987.
Stace W. T., *The Philosophy of Hegel,* Nueva York, Dover, 1955 (1924).
Stahl F. J., cf. Riedel.
Steinbüchel Theodor, Das *Grundproblem der Hegelschen Philosophie,* Bonn, Hanstein, 1933.
Stoebe H. J., *Gut und Bóse in der jahwistischen Quelle des Pentateuch.* Zeitschrift für Alttestamentliche Wissenschaft 65 (1953), pp. 188-204.
Taylor E. F. y J. A Wheeler., *Spacetime Physics,* San Francisco, Freeman, 1966.
Teevan R. C. y R. C Birney., Teorías *sobre motivación del aprendizaje,* México, Trillas, 1976.
Theunissen Michael, *Hegels Lehre vom absoluten Geist als theologischpolitischer Traktat,* Berlín, Gruyter, 1970
Weil Eric, *Hegel et l'état,* París, Vrin, 1974. Von Weizsäcker C. F., cf. Bastin.

Westermann Claus, *Génesis 1-11.* Wiss, Darmstadt, Buchgesellschaft, 1976.
White Leslie A., *La Ciencia de la cultura,* tr. Gerardo Steenks, Buenos Aires, Paidós, 1964
______, cf. Rossi.
Zerwick Max, *Analysis philologica Novi Testamenti graeci,* Romae, PIB, 1960.
______, *Graecitas bíblica,* Roma, PIB, 1960.
Zubiri Xavier, *Sobre la esencia,* Madrid, Sociedad de Estudios y Publicaciones, 1963.

Hegel's works

Ästh *Ásthetik* (2 vols.), Reclam, Stuttgart, 1971, Frankfurt, BS *Berliner Schriften.* Suhrkamp, 1970.
DG *Vorlesungen über die Beweise vom Dasein Gottes,* Hamburgo, Himmeleber, 1966.
EGP *Einleitungin die Geschichte der Philosophie,* Hamburgo, Meiner, 1966.
EPW *Enzyklopädie der philosophischen Wissenschaften,* Hamburgo, Meiner, 1969.
FS *Frühe Schriften,* Frankfurt, Suhrkamp, 1974.
FR *Filosofía Real,* tr. José Ma. Ripalda, México, FCE, 1984
GP *Geschichte der Philosophie,* Frankfurt, Suhrkamp, 1975, 3 vols.
JLMN *Jenenser Logik Metaphysik und Naturphilosophie,* Hamburgo, Meiner, 1967.
JS *Jenaer Schriften,* Frankfurt, Suhrkamp, 1974.
NH *Nümbergund Heidelberg Schriften,* Frankfurt, Suhrkamp, 1975.
PG *Phänomenologie des Geistes,* Hamburgo, Meiner, 1952.
PR *Vorlesungen über die Phüosophie der Religión,* Hamburgo, Meiner, 1974, 2 vols.
Rph *Grundlinien der Philosophie des Rechts,* Frankfurt, Ullstein, 1972.
VG *Die Vemunft in der Geschichte,* Hamburgo, Meiner, 1980.
WG *Vorlesungen über die Philosophie der Weltgeschichte,* Hamburgo, Meiner, 1976.
WL *Wissenschaft der Logik,* Hamburgo, Meiner, 1963.

Other

EB *Encyclopaedia Britannica.* The University of Chicago, 1986,32 vols.
LTK *Lexikon für Theologie und Kirche.* Hender, Freiburg, 1957-1965.
MEW Marx *Engels Werke.* Dietz, Berlín, 1973-1974, 43 vols.

Quotes from HEGEL'S Phenomenology of Spirit and Hegel's Science of Logic are the translator AV Miller.

DAEDALUS
Europäisches Denken in deutscher Philosophie

Herausgegeben von Volker Caysa, Hans-Martin Gerlach †, Leon Miodonski, Friedrich Tomberg und Klaus Vieweg

Band 1 Thomas Grüning / Klaus Vieweg: Hegel - Vision und Konstruktion einer Vernunftgeschichte der Freiheit. 1991.

Band 2 Eva-Maria Tschurenev: Kant und Burke: Ästhetik als Theorie des Gemeinsinns. 1992.

Band 3 János Rathmann: Historizität in der Deutschen Aufklärung. 1993.

Band 4 Lars Lambrecht / Eva-Maria Tschurenev (Hrsg.): Geschichtliche Welt und menschliches Wesen. Beiträge zum Bedenken der conditio humana in der europäischen Geistesgeschichte. 1994.

Band 5 Hans-Martin Gerlach / Hans Rainer Sepp (Hrsg.): Husserl in Halle. Spurensuche im Anfang der Phänomenologie. 1994.

Band 6 Volker Caysa: Das Seyn entwerfen. Die negative Metaphysik Martin Heideggers. 1994.

Band 7 Jürgen Friedrich / Bernd Westermann (Hrsg.): Unter offenem Horizont. Anthropologie nach Helmuth Plessner. Mit einem Geleitwort von Dietrich Goldschmidt. 1995.

Band 8 Jaromir Brejdak: Philosophia crucis. Heideggers Beschäftigung mit dem Apostel Paulus. 1996.

Band 9 Volker Caysa (Hrsg.): Auf der Suche nach dem Citoyen. Konzepte der Citoyenität. 1997.

Band 10 Thies Schröder: Natur als Idee. Begründung und Aktualität des Naturbegriffs unter kritischer Berücksichtigung der Naturphilosophie Schellings. 1997.

Band 11 Jan Rachold: Die aufklärerische Vernunft im Spannungsfeld zwischen rationalistisch-metaphysischer und politisch-sozialer Deutung. Eine Studie zur Philosophie der deutschen Aufklärung (Wolff, Abbt, Feder, Meiners, Weishaupt). 1999.

Band 12 Hans-Jürgen Arendt: Gustav Theodor Fechner. Ein deutscher Naturwissenschaftler und Philosoph im 19. Jahrhundert. 1999.

Band 13 Reinhard Zecher: Das Ziel der Einheit. Verwirklichung einer Idee oder Ergebnis eines Selbstorganisationsprozesses? Schellings spekulative "Naturphilosophie" und die Evolutionstheorie: Gesprächspartner in der Frage nach einer Rückbesinnung auf die Naturphilosophie? 2000.

Band 14 Marion George: Die Geburt der Hermeneutik aus dem Geist der Politik. Diltheys Begründung der Philosophie als pragmatische Ordnungsmacht der Moderne. 2002.

Band 15 Christian Rabanus: Praktische Phänomenologie. Jan Patockas Revision der Phänomenologie Edmund Husserls. 2002.

Band 16 Hans-Martin Gerlach / Andreas Hütig / Oliver Immel (Hrsg.): Symbol, Existenz, Lebenswelt. Kulturphilosophische Zugänge zur Interkulturalität. 2004.

Band 17 Stefan Groß: Die Metaphysik Karl Christian Friedrich Krauses in ihrem Verhältnis zu Religion, Ethik und Ästhetik. 2008.

Band 18 Volker Caysa / Bartłomiej Kozera / Justus H. Ulbricht (Hrsg.): Kultur – Nation – Europa. Nationalkulturelle Identitäten auf einem imaginären Kontinent. 2008.

Band 19 José Porfirio Miranda de la Parra: Hegel was right. The Myth of the Empirical Sciences. Translation by Eduardo Charpenel Elorduy. 2011.

www.peterlang.de

www.ingramcontent.com/pod-product-compliance
Ingram Content Group UK Ltd.
Pitfield, Milton Keynes, MK11 3LW, UK
UKHW021831100926
8556IPUK00004B/271

9 783631 618899